Chemistry for the Logic Stage

Teacher Guide

Chemistry for the Logic Stage Teacher Guide

Second Edition (First Printing 2020)
Copyright @ Elemental Science, Inc.
Email: support@elementalscience.com

ISBN# 978-1-953490-03-2

Printed in the USA for worldwide distribution

For more copies write to:
Elemental Science
PO Box 79
Niceville, FL 32588
support@elementalscience.com

Copyright Policy

Quick Start Guide

In a Nutshell

Learn about the periodic table, matter, solutions, chemical reactions, acids, bases, the chemistry of life, and the chemistry of industry through the following:

- ✓ Gathering information through reading the main spines.

- ✓ Doing hands-on science through experiments and projects.

- ✓ Keeping a record of what the students have learned.

See pp. 17-18 for a complete list of the topics explored in this program.

What You Need

In addition to this guide, you will need the following:

1. A guide for the students. (You can purchase the *Chemistry for the Logic Stage Student Guide* to have it all laid out for you or just buy a composition book.)

2. The two spines:
   - 📖 *Usborne Science Encyclopedia, 2015 Edition* (USE)
   - 📖 *Usborne Illustrated Dictionary of Science, 2012 Edition* (UIDS)

   You can also purchase the *Kingfisher Science Encyclopedia, 2017 Edition* (KSE) for optional reading assignments. Head to the page below to get links to these books:

   💻 https://elementalscience.com/blogs/resources/cls

3. The experiment supplies (See a full list starting on pg. 19 or save yourself the time and purchase the *Chemistry for the Logic Stage Experiment Kit*.)

How It Works

Each week you will . . .

- ☞ Guide the students as they do an experiment using the directions on the Student Assignment Sheet—this is in this guide as well as the student guide. The results and an explanation of the experiment are part of the additional information in this guide.

- ☞ Assign the reading, and when the students are finished, you will discuss what they read using the questions and answers in this guide.

- ☞ Assign appropriate written work—a list of facts, an outline, or a report.

- ☞ Assign vocabulary, memory work, and dates to add to a timeline.

You can add to their learning experience by also doing the additional activities suggested in this guide. For a more detailed explanation of the components of your week, we highly recommend reading the introduction starting on pg. 7 of this guide.

Chemistry for the Logic Stage Teacher Guide
Table of Contents

6

Appendix

Templates 265

Chemistry for the Logic Stage
Introduction

In *Success in Science: A Manual for Excellence in Science Education*, we state that the middle school student is "a bucket full of unorganized information that needs to be filed away and stored in a cabinet."[1] The goals of science instruction at the logic level are to begin to train the students' brain to think analytically about the facts of science, to familiarize the students with the basics of the scientific method through inquiry-based techniques and to continue to feed the students with information about the world around them. *Chemistry for the Logic Stage* integrates the above goals using the Classic Method of middle school science instruction as suggested in our book. This method is loosely based on the ideas for classical science education that are laid out in *The Well-Trained Mind: A Guide to Classical Education at Home* by Jessie Wise and Susan Wise Bauer.

This guide includes the four basic components of middle school science instruction as explained in *Success in Science*.

1. **Hands-on Inquiry** — Middle school students need to see real-life science, to build their problem solving skills and to practice using the basics of the scientific method. This can be done through experiments or nature studies. In this guide, the weekly experiments fulfill this section of middle school science instruction.
2. **Information** — Middle school students need to continue to build their knowledge base along with learning how to organize and store the information they are studying. The information component is an integral part of this process. In this guide, the reading assignments, vocabulary and sketches contain all of the necessary pieces of this aspect of middle school science instruction.
3. **Writing** — The purpose of the writing component is to teach the students how to process and organize information. You want them to be able to read a passage, pull out the main ideas and communicate them to you in their own words. The assigned outlines or reports in this guide give you the tools you need to teach this basic component to your student.
4. **The Science Project** — Once a year, all middle school student should complete a science project. Their project should work through the scientific method from start to finish on a basic level, meaning that their question should be relatively easy to answer. The science fair project, scheduled as a part of unit seven fulfills the requirements of this component.

Chemistry for the Logic Stage also includes the two optional components of middle school science instruction as explained in *Success in Science*.

1. **Around the Web** — Middle school students should gain some experience with researching on the Internet. So for this optional component, the students should, under your supervision, search the Internet for websites, YouTube videos, virtual tours and activities that relate to what they are studying. In this guide, the Want More lessons recommend specific sites and activities for you to use.
2. **Quizzes or Tests** — During the middle school years it is not absolutely necessary that you give quizzes or tests to the students. However, if you want to familiarize them with

[1] Bradley R. Hudson & Paige Hudson, Success in Science: A Manual for Excellence in Science Education, (Elemental Science, 2012) 52

test-taking skills, we suggest that you give quizzes or tests that will set the students up for success. With that in mind, we have included optional tests for you to use with each unit.

My goal in writing this curriculum is to provide you with the tools to explore the field of chemistry while teaching the basics of the scientific method. During the years, your students will work on their observation skills, learn to think critically about the information they are studying and practice working independently. *Chemistry for the Logic Stage* is intended to be used with seventh through eighth grade students.

What this guide contains in a nutshell

This guide includes the weekly student assignment sheets, all the sketches pre-labeled for you and discussion questions to help you guide your discussion time. This guide also contains information for each experiment, including the expected results and an explanation of those results. There is a list of additional activities that you can choose to assign for each week. Finally, this guide includes possible schedules for you to use as you guide your students through *Chemistry for the Logic Stage*.

What the Student Guide contains

The Student Guide, which is sold separately, is designed to encourage independence in your students as they complete *Chemistry for the Logic Stage*. The Student Guide contains all the student assignment sheets, pre-drawn sketches ready for labeling, experiment pages and blank report pages. The guide also includes blank date sheets as well as all the sheets they will need for the Science Fair Project. In short, the Student Guide contains all the pages your students will need and it is essential for successfully completing this program.

Student Assignment Sheets

This Teacher Guide contains a copy of each of the student assignment sheets that are in the Student Guide. This way you can stay on top of what your students are studying. Each of the student assignment sheets contains the following:

✓ **Experiment**

Each week will revolve around a weekly topic that it to be studied. Your student will be assigned an experiment that poses a question related to the topic. Each of these experiments will walk your students through the scientific method (see the Appendix pg. 249 for a brief explanation of the scientific method). In a nutshell, the scientific method trains the brain to examine and observe before making a statement of fact. It will teach your student to look at all the facts and results before drawing a conclusion. If this sounds intimidating, it's not. You are simply teaching your students to take the time to discover the answer to a given problem by using the knowledge they have and the things they observe during an experiment.

Each week, the student assignment sheet will contain a list of the materials needed and the instructions to complete the experiment. The student guide contains an experiment sheet for your students to fill out. Each experiment sheet contains an introduction that is followed by a list of materials, a hypothesis, a procedure, an observation and a conclusion section.

The introduction will give your students specific background information for the experiment. In the hypothesis section, they will predict the answer to the question posed in the lab. In the materials listed section, your students will fill out what they will use to complete the experiment. In the procedure section, they will recount step by step what was done during their experiment, so that someone else could read their report and replicate their experiment. In the observation section, your students will write what they saw. Finally, in the conclusion section they will write whether or not their hypothesis was correct and share any additional information they have learned from the experiment. If the students' hypothesis was not correct, discuss why and have them include that on their experiment sheet.

Vocabulary & Memory Work

Throughout the year, the students will be assigned vocabulary for each week. They will need to write out the definitions for each word on the Unit Vocabulary Sheet found in the Student Guide on the week that they are assigned. You may want to have your students also make flash cards to help them work on memorizing the words. This year, the students will memorize the elements of the periodic table along with specific information relating to each unit. There is a complete listing of the vocabulary words and memory work for each unit on the unit overview sheet in this guide along with a glossary and a list of the memory work in the Student Guide.

Sketch

Each week the students will be assigned a sketch to complete and label. The Student Guide contains an unlabeled sketch for them to use. They will color the sketch, label it and give it a title according to the directions on the Student Assignment Sheet. The information they need will be in their reading, but the sketch is not always identical to the pictures found in the encyclopedia. So, these sketch assignments should make the student think. This guide contains a completed sketch for you to use when checking their work.

Writing

Each week the students will be assigned pages to read from the spine text, the *Usborne Science Encyclopedia* or the *Usborne Illustrated Dictionary of Science*. Have them read the assigned pages and discuss what they have read with you. After you have finished reading and discussing the information, you have three options for your students' written assignment:

> *Option 1: Have the students write an outline from the spine text*
> A typical seventh grader completing this program should be expected to write a two to three level outline for the pages assigned for the week. This outline should include the main point from each paragraph on the page as well as several supporting and sub supporting points;

> *Option 2: Have the students write a narrative summary from the spine text*
> A typical seventh grader completing this program should be expected to write a three to six paragraph summary (or about a page) about what they have read in the spine text;

> *Option 3: Have the students write both an outline and a written report*
> First, have the students read the assigned pages in the spine text. Then, have them write a two to three level outline for the assigned pages. Next, have the students do

some additional research reading on the topic from one or more of the suggested reference books listed below. Each topic will have pages assigned from these reference books for their research. In addition to the main spines, the following encyclopedia is scheduled to be used as an additional reference book:

📖 *The Kingfisher Science Encyclopedia, 2017 Edition* (KSE)

Once the students complete the additional research reading, have them write a report of three to four paragraphs in length, detailing what they have learned from their research reading.

Your writing goal for middle school students is to have them write something (narrative summary, outline or list of facts) every day you do school, either in science or in another subject. So, the writing option you choose for this curriculum will depend on the writing the students are already doing in their other subjects.

When evaluating the students' report, make sure that the information they have shared is accurate and that it has been presented in a grammatically correct form (i.e., look for spelling mistakes, run-on sentences and paragraph form). In the Student Guide, there are two blank lined sheets for the students to use when writing their outlines and/or summaries. If you are having the students type their report, have them glue a copy of it into their Student Guide.

🕐 **Dates**

Each week the dates of important discoveries within the topic and the dates from the readings are given on the student assignment sheet. The students will enter these dates onto one of their date sheets. The date sheets are divided into the four time periods as laid out in *The Well-Trained Mind* by Susan Wise Bauer and Jessie Wise (Ancients, Medieval-Early Renaissance, Late Renaissance-Early Modern, and Modern). Completed date sheets are available for you to use in the appendix of this guide on pp. 246-248.

Schedules

Chemistry for the Logic Stage is designed to take up to 5 hours per week. You and your students can choose whether to complete the work over five days or over two days. Each week I have included two scheduling options for you to use as you lead them through this program. They are meant to be guides, so feel free to change the order to better fit the needs of your students. I also recommend that you begin to let them be in charge of choosing how many days they would like to do science as this will help to begin to foster independence in their school work.

Additional Information Section

The Additional Information Section includes tools that you will find helpful as you guide the students through this study. It is only found in the Teacher Guide, and it contains the following:

☞ **Experiment Information**

Each week, the Additional Information Section includes the expected experiment results and an explanation of those results for you to use with the students. When possible, you will also find suggestions on how to expand the experiment in the Take if Further section.

🍀 **Discussion Questions**

Each week the Additional Information Section includes possible discussion questions from the main reading assignment, along with the answers. These are designed to aid you in leading the discussion time with the students. I recommend that you encourage them to answer in complete sentences, as this will help them organize their thoughts for writing their outline or report. I have also included a list of the discussion questions without the answers at the end of each unit's material in this guide. This is so you can give them to your students ahead of time, if you desire, or you can use them to review for the unit test. If they are already writing outlines or lists of facts, you do not need to have them write out the answers to the discussion questions before hand as there is plenty of writing required in this program already.

⟶ Want More

Each week, the Additional Information Section includes a list of activities under the Want More section. ***These activities are totally optional.*** The Want More activities are designed to explore the science on a deeper level by researching specific topics or through additional projects to do. The students do not have this information in their guide, so it is up to you whether or not to assign these.

☑ Sketch

Each week, the Additional Information Section includes copies of the sketches that have been labeled. These are included in this guide for you to use as you correct the students' work.

Tests

The students will be completing a lot of work each week that will help you to assess what they are learning, so testing is not absolutely necessary. However, I have included end of unit tests that you can use if you feel the need to do so. The tests and the answers are included after the material for each unit in this guide. You can choose to give the tests orally or copy them for the students to fill out.

What a Typical Two-day Schedule Looks Like

A typical two-day schedule will take one-and-a-half to two hours per day. Here is a breakdown of how a normal two-day week would work using week three:

- **Day 1: *Define the vocabulary, record the dates, do the experiment, and complete the experiment sheet***
  Begin day 1 by having the students do the "Can I transfer metal atoms?" experiment. Have them read the introduction and perform the experiment using the directions provided. Next, have them record their observations and results. After they discuss their results with you, have them write a conclusion for their experiment. Finish the day by having them look up and define "metal" using the glossary in the Student Guide and add the dates to their date sheets.

- **Day 2: *Read the assigned pages, discuss together, prepare an outline, or narrative summary and complete the sketch***
  Begin by having the students read pp. 168,170, 172-173 in the *Usborne Illustrated Dictionary of Science*. Then, using the questions provided, discuss what they have read. Next, have them complete the sketch using the directions on the Student Assignment Sheet. Finally, have them write an outline or narrative summary. Here is what that could look like:

Alkali, Alkaline Earth, and Transition Metals

Alkali metals are metals that react with water to form alkaline solutions. They are typically soft, silver-white metals. The further down the group, the more reactive the alkali metal elements are.

Alkali metals are used in a variety of products. Lithium is used in welding flux. Sodium is used as a coolant in nuclear power plants. Potassium is used to make soap. Rubidium is used to make a special type of glass. Cesium is used in photocells. Francium, the final alkali metal does not have a known stable isotope.

Alkaline earth metals are reactive metals. With the exception of Beryllium, which is a hard, white metal, the alkaline earth metals are soft, silver-white metals. The further down the group, the more reactive the alkaline earth metal elements are.

Alkaline earth metals are found in many places and have many uses. Beryllium is used in corrosion-resistant alloys. Magnesium is found in rocks and seawater. Calcium is found in chalk, milk, and bones. Strontium is used in fireworks. Barium is used in medicine. And, radium is used to treat cancer.

Transition metals are all hard, tough, shiny, and malleable metals. These metals also typically conduct heat and electricity. The transition metals also tend to have high melting points, boiling points, and densities. The inner transition metals are rare and often unstable.

Transition Metals are used in a variety of ways. Many transition metals are used in alloys. Many transition metals are used as catalysts. Some transition metals are used in electroplating.

What a Typical Five-day Schedule Looks Like

A typical five-day schedule will take forty-five minutes to one hour per day. Here is a breakdown of how a normal five-day week would work using week three...

🦶 **Day 1: *Do the experiment and complete the experiment sheet***
Begin day 1 by having the students do the "Can I transfer metal atoms?" experiment. Have them read the introduction and perform the experiment using the directions provided. Next, have them record their observations and results, discuss their results with you and then write a conclusion for their experiment;

🦶 **Day 2: *Read the assigned pages, discuss together, and write an outline or list of facts***
Begin by having the students read pp. 168,170, 172-173 in the *Usborne Illustrated Dictionary of Science* and discuss what they have read using the provided questions. Then, have the students write a two to three level outline, and complete the sketch using the directions on the Student Assignment Sheet. Here's a sample outline:

Alkali Metals

I. Alkali metals include the Group 1 elements on the periodic table.
   A. All metals are metals that react with water to form alkaline solutions.

     i. The further down the group, the more reactive the alkali metal elements are.

  B. Alkali metals are typically soft, silver-white metals.

II. Alkali metals are used in a variety of products.

  A. Lithium is used in welding flux.

  B. Sodium is used as a coolant in nuclear power plants.

  C. Potassium is used to make soap.

  D. Rubidium is used to make a special type of glass.

  E. Cesium is used in photocells.

Alkaline Earth Metals

I. Alkaline earth metals include the Group 2 elements on the periodic table.

  A. Alkaline earth metals are reactive metals.

     i. The further down the group, the more reactive the alkaline earth metal elements are.

  B. With the exception of Beryllium, which is a hard, white metal, the alkaline earth metals are soft, silver-white metals.

II. Alkaline earth metals are found in many places and have many uses.

  A. Beryllium is used in corrosion-resistant alloys.

  B. Magnesium is found in rocks and seawater.

  C. Calcium is found in chalk, milk and bones.

  D. Strontium is used in fireworks.

  E. Barium is used in medicine.

  F. Radium is used to treat cancer.

Transition Metals

I. Transition metals include the center block of elements on the periodic table.

  A. Transition metals are all hard, tough, shiny, and malleable metals.

  B. These metals also typically conduct heat and electricity.

  C. The transition metals also tend to have high melting points, boiling points, and densities.

  D. The inner transition metals are rare and often unstable.

II. Transition Metals are used in a variety of ways.

  A. Many transition metals are used in alloys.

  B. Many transition metals are used as catalysts.

  C. Some transition metals are used in electroplating.

↳ **Day 3: *Record the dates, define the vocabulary, and complete the sketch***
Begin by having the students look up and define "metal" using the glossary in the Student Guide and add the dates to their date sheets. Then, have them complete the sketch using the directions on the Student Assignment Sheet;

↳ **Day 4: *Read from the additional reading assignments and prepare a written report***
Begin by having the students read "Metals" from *KSE* pp. 183, "Copper" from *KSE* pg.

199, or "Metals" from *USE* pp. 30-31. Then, have the students use their outline along with what they have just read to write a 3 to 5 paragraph summary of what they have learned;

ᴥ **Day 5: *Complete one of the Want More activities***
Have the students make several Element Trading Cards or have them watch the video about alkali metals. You could also have them read about a scientist from the field of chemistry.

The Science Fair Project

I have scheduled time for the students to complete a science fair project during unit seven. Janice VanCleave's *A+ Science Fair Projects* and Janice VanCleave's *A+ Projects in Chemistry: Winning Experiments for Science Fairs and Extra Credit* are excellent resources for choosing project topics within the field of chemistry. You can call your local school system to see if it allows homeschooled students to participate in the local school science fair or get information on national science fairs from them. Another option would be to have your students present their project in front of a group of friends and family.

How to Include Your Younger Students

I recognize that many homeschool families have a range of different student ages. If you wish to have all your students studying the topic of chemistry you have two options for your elementary students when using this program with your middle school students:

ᴥ ***Option 1: Have your younger students use Chemistry for the Grammar Stage***
I recommend this option if your younger students are in the second through fourth grade and/or your older students are ready for some independence. The units in *Chemistry for the Grammar Stage* will not match up with the units in *Chemistry for the Logic Stage*, so you will need to do each program as written;

ᴥ ***Option 2: Have your younger students use Chemistry for the Logic Stage along with your older students***
I recommend this option if your younger students are in the fourth through sixth grade and/or your older students are not ready to work independently. However, you will need to adjust the work load for your younger students. Here are some suggestions on how to do that:
✓ Have them watch and observe the experiments;
✓ Add in some picture books from the library for each of the topics;
✓ Read the reading assignments to them and have them narrate them back to you;
✓ Let them color the sketches and then tell them how to label them.
As for the reading assignments, you may find that the spines scheduled are too much for your younger students. If so, you can read to them out of the *DK Smithsonian Science: A Visual Encyclopedia*. I have included a chart coordinating this resource in the Appendix on pp.250-252.

Helpful Articles

Our goal is to provide you with the information you need to be successful in your quest to educate your students in the sciences at home. This is the main reason we share tips and tools for homeschool science education on our blogs. As you prepare to guide your students through this program, you may find the following articles helpful:

- *Classical Science Curriculum for the Logic Stage Student* – This article explains the goals of logic stage science and demonstrates how the classical educator can utilize the tools they have at their disposal to reach these goals.
  - http://elementalblogging.com/classical-science-curriculum-logic/
- *Scientific Demonstrations vs. Experiments* – This article shares information about these two types of scientific tests and points out how to employ scientific demonstrations or experiments in your homeschool.
  - https://elementalscience.com/blogs/news/89905795-scientific-demonstrations-or-experiments
- *Writing in Homeschool Science: The Middle School Years and Beyond* – This podcast (and video) explains the goals of writing for logic stage science.
  - https://elementalscience.com/blogs/podcast/episode-13
- *A Simple Explanation of the Scientific Method* – This article details the steps of the scientific method, along with why it is so important to teach.
  - https://elementalscience.com/blogs/news/simple-explanation-of-the-scientific-method/
- *3 Tips to Encourage Independent Learning* – This podcast gives you tips to help your students make the move from dependent to independent learning.
  - https://elementalscience.com/blogs/podcast/87

Additional Resources

The following page contains quick links to the activities suggested in this guide along with several helpful downloads:

- https://elementalscience.com/blogs/resources/cls

Final Thoughts

If you find that this program contains too much work, please tailor it to the needs of your student. As the author and publisher of this curriculum I encourage you to contact me with any questions or problems that you might have concerning *Chemistry for the Logic Stage* at support@elementalscience.com. I will be more than happy to answer them as soon as I am able. I hope that you and your student enjoy *Chemistry for the Logic Stage*!

Safety Advisory

Many of the experiments in this book use boiling water or open flames. We recommend that your students use safety glasses and protective gear with each experiment to prevent accidents. Do not allow your students to perform any of the experiments marked " ☺ **CAUTION** " on their own.

Units of Measurement

What are the two measuring systems?

- **The Standard or Standard American Engineering (SAE) System** — This system is used mainly in the United States and it uses units like inches, pounds and gallons. It was derived from an early English measuring system that has its roots in the Roman system of measurements.
- **The Metric System** — This system is used in most of the world and it uses units like meters, grams and liters. The system is base 10 and their names are formed with prefixes. It was derived from one of the early French measuring systems.

In the US, the standard system of units are more widely used on consumer products and in industrial manufacturing, while the metric system is more widely used in science, medicine and government. Since this program has been published in the US, I have used the standard system of measurement throughout for familiarity. However, because I believe that it is important for our students to be familiar with both systems, I have included metric measurements in parentheses.

What about converting units?

Every student should know how to convert measurements inside of a given measuring system, such as knowing how to convert grams to kilograms or ounces to pounds. Normally, these conversion factors are taught as a part of your math program. However, I also recommend that you have your students memorize several basic conversion factors between the two systems. Here is a list of factors that the students should try to memorize:

- **Pounds to Kilograms:** 1 lb = 2.2 kg
- **Ounces to Grams:** 1 oz = 28.3 g
- **Gallons to Liters:** 1 gal = 3.785 L
- **Cups to Milliliters:** 1 c = 240 mL
- **Miles to Kilometers:** 1 mi = 1.61 km
- **Feet to Meters:** 1 ft = 0.305 m
- **Inches to Centimeters:** 1 in = 2.54 cm

With the global flow of information that occurs these days, it is very important for students to learn these most basic conversion factors. To learn more about the importance of units of measurement in science, read the following blog post:

🖥 https://elementalscience.com/blogs/science-activities/units-of-measurement

Sequence of Study

Building Blocks of Chemistry (9 weeks)

Unit 1: The Periodic Table (5 Weeks)
- ✓ Elements, Atoms and Ions
- ✓ The Periodic Table
- ✓ Metals
- ✓ The Inbetweens
- ✓ Halogens and Noble Gases

Unit 2: Matter (4 Weeks)
- ✓ States of Matter
- ✓ Properties of Matter
- ✓ Kinetic Theory and Gases
- ✓ Crystals

Principles in Chemistry (14 weeks)

Unit 3: Solutions (4 Weeks)
- ✓ Compounds and Mixtures
- ✓ Solutions
- ✓ Separating Mixtures
- ✓ Electrolysis

Unit 4: Chemical Reactions (6 Weeks)
- ✓ Chemical Bonding
- ✓ Chemical Reactions
- ✓ Reactivity
- ✓ Catalysts
- ✓ Oxidation and Reduction

Unit 5: Acids and Bases (4 weeks)
- ✓ Acids
- ✓ Bases
- ✓ Measuring Acidity (pH)
- ✓ Neutralization and Salts

Applications for Chemistry (12 weeks)

Unit 6: Chemistry of Life (4 Weeks)
- ✓ Organic Chemistry

✓ Enzymes
✓ Chemistry of Food
✓ Fermentation

Unit 7: Chemistry of Industry (8 weeks)
✓ Soaps and Detergents
✓ Alkanes and Alkenes
✓ Homologous Groups
✓ Petrochemicals
✓ Polymers and Plastics
✓ Iron and Alloys
✓ Radioactivity
✓ Pollution

Note—This unit also contains a science fair project and a scientist biography project for the students to complete.

Materials Listed by Week

Building Blocks of Chemistry

Unit 1: The Periodic Table

Week	Materials
1	Heavy cream, Milk, Sugar, Vanilla, 1 small and 1 large zip-locking plastic bag, Crushed ice, Dish towel or oven mitt, Rock salt
2	*No supplies needed.*
3	Vinegar, Salt, 6 pennies, Glass cup, 2 iron nails
4	Magnet, Materials for circuit (Flashlight bulb, Copper wire, D battery, Electrical tape), Hammer, Paper folded into a small square, Metal paperclip (not plastic coated), Aluminum foil, CD, Safety glasses
5	Element cards (homemade or purchased)

Unit 2: Matter

Week	Materials
6	Cup, Ice Cubes, Pot, Thermometer
7	4 clear cups, Eye dropper, Table salt, Food coloring, Water
8	2 cups, Apple juice, Timer, Partner
9	String, Wide mouthed jar, Pencil, Pipe cleaners, Water, Borax, Scissors

Principles in Chemistry

Unit 3: Solutions

Week	Materials
10	Bag of multi-colored marshmallows, Toothpicks
11	5 clear cups (or beakers), 5 plastic spoons, Sugar, Salt, Baking powder, Flour, Cornstarch, Water, Vegetable oil, Tablespoon
12	Coffee filters, Markers, Alcohol, Coffee can or wide-mouthed jar, Rubber bands, Eyedropper
13	Distilled water, 2 test tubes, Salt, Glass cup, 2 Alligator clips, Covered copper wire, 6-volt Lantern battery, Permanent marker

Unit 4: Chemical Reactions

Week	Materials
14	Cake frosting, Red and yellow bite-sized candies
15	Yeast, Hydrogen peroxide, Epsom salts, Water, 2 cups, 2 thermometers
16	Baking soda, Chalk, Iron nail (non-coated), Copper penny, White vinegar, 4 cups

Week	Materials
17	2 potatoes, Pot, Water, Oven mitt, Large Slotted Spoon
18	Carbonated water, Sugar, 2 cups
19	Steel wool, Vinegar, Jar with lid, Ammonia, Hydrogen peroxide,

Unit 5: Acids and Bases

Week	Materials
20	Cranberry juice, Lemon juice, Baking soda, Clear cup
21	6 cups, Red cabbage solution, Water, Vinegar, Baking soda, Sprite, Ammonia, Lemon Juice, Eye dropper
22	Lemon, Tomato, Saliva, Milk, Bleach, Toothpaste, Liquid Dish Soap, pH paper, Gloves
23	Vinegar, Ammonia, Red cabbage solution, Water, Safety glasses

Applications for Chemistry

Unit 6: Chemistry of Life

Week	Materials
24	Sugar, Salt, Candle, 2 metal spoons, Hot mitt
25	2 slices of bread, Water, Saliva, 2 plastic bags
26	Benedict's solution, Iodine solution, Several different types of food for testing (such as a hard-boiled egg, bread, potato, pasta, yogurt, cookies or cheese), Eyedropper, Small plastic cups, Safety glasses
27	Yeast, Water, Sugar, 3 bottles, 3 balloons, Instant read thermometer, Pot, Hot mitt

Unit 7: Chemistry of Industry

Week	Materials
28	Powdered detergent, Liquid soap, 2 large cups, 2 small cups, 2 bowls, pH paper, Vegetable oil, Dirt, Ketchup, Plaster of Paris, Water, Straw, Old T-shirt fabric
29-36	Materials will vary depending on the Science Fair Project that your student has chosen to do.
33-35	*No supplies needed.*

Chemistry: Unit 1

The Periodic Table

Unit 1: The Periodic Table
Overview of Study

Sequence of Study

Week 1: Atoms
Week 2: The Periodic Table
Week 3: Metals
Week 4: The Inbetweens
Week 5: Halogen and Noble Gases

Materials by Week

Week	Materials
1	Heavy cream, Milk, Sugar, Vanilla, 1 small and 1 large zip-locking plastic bag, Crushed ice, Dish towel or oven mitt, Rock salt
2	*No supplies needed.*
3	Vinegar, Salt, 6 pennies, Glass cup, 2 iron nails
4	Magnet, Materials for circuit (Flashlight bulb, Copper wire, D battery, Electrical tape), Hammer, Paper folded into a small square, Metal paperclip (not plastic coated), Aluminum foil, CD, Safety glasses
5	Element cards (homemade or purchased)

Vocabulary for the Unit

1. **Atom** – The smallest particle of an element that retains the chemical properties of the element.
2. **Electron Shell** – A region around the nucleus of an atom where a specific number of electrons can exist.
3. **Element** – A substance made up of one type of atom, which cannot be broken down by chemical reaction to form a simpler substance.
4. **Compound** – A substance made up of two or more different elements that are chemically joined in fixed proportions.
5. **Period** – A set of elements that have the same number of electron shells, shown as rows in the periodic table.
6. **Group** – A column of elements in the periodic table that have similar properties, electron configurations and valencies.
7. **Atomic Number** – The number of protons in the nucleus of an atom.
8. **Atomic Mass** – The average mass number of the atoms in a sample of an element.
9. **Metal** – The largest class of elements, usually they are shiny and solid at room temperature.
10. **Poor Metal** – A group of metals that are soft and weak.
11. **Semimetal** – A group of elements that have characteristics of both metals and nonmetals.

12. **Malleable** – A characteristic of a metal that means it is bendable and easily shaped.
13. **Semiconductor** – A substance that only conducts electricity under certain conditions.
14. **Nonmetal** – A class of elements that typically forms negative ions; they are usually dull solids or gases.

Memory Work for the Unit

The Elements of the Periodic Table – The following elements will be memorized in this unit:

- ✓ 1-H-Hydrogen
- ✓ 2-He-Helium
- ✓ 3-Li-Lithium
- ✓ 4-Be-Beryllium
- ✓ 5-B-Boron
- ✓ 6-C-Carbon
- ✓ 7-N-Nitrogen
- ✓ 8-O-Oxygen
- ✓ 9-F-Fluorine
- ✓ 10-Ne-Neon
- ✓ 11-Na-Sodium
- ✓ 12-Mg-Magnesium
- ✓ 13-Al-Aluminum
- ✓ 14-Si-Silicon
- ✓ 15-P-Phosphorus
- ✓ 16-S-Sulfur
- ✓ 17-Cl-Chlorine
- ✓ 18-Ar-Argon
- ✓ 19-K-Potassium
- ✓ 20-Ca-Calcium

Student Assignment Sheet Week 1
Atoms

Experiment: Making Ice Cream

Materials:

- ✓ Heavy cream
- ✓ Milk
- ✓ Sugar
- ✓ Vanilla
- ✓ Crushed ice

- ✓ 1 small & 1 large zip-locking plastic bag
- ✓ Dish towel or oven mitt
- ✓ Rock salt

Procedure:

1. Mix together the ½ cup cream (120 mL), 1 cup milk (240 mL), ½ cup sugar (225 g) and 1 tsp vanilla (5 mL) in the small zip-locking plastic bag. Then, add the crushed ice and rock salt to the large zip-locking plastic bag. (**Note**—*Make sure that the air is removed from both bags and they are sealed tightly.*)
2. Place the smaller bag with the cream mixture inside the larger bag with the ice. Cover the outside of the bag with the dish towel (or put on the oven mitt) and massage or shake the bag until the cream mixture has frozen. It should take about 5 to 10 minutes.
3. Take the smaller bag out of the larger bag, wipe off the salt water, open, eat and enjoy!

Vocabulary & Memory Work

- ☐ Vocabulary: atom, electron shell, element, compound
- ☐ Memory Work—This year you will be memorizing the elements that make up the periodic table, along with their atomic number and symbol. This week, work on the following elements:
  - ✓ 1-H-Hydrogen, 2-He-Helium, 3-Li-Lithium, 4-Be-Beryllium

Sketch: Parts of an Atom

- ▣ Label the following: electron orbit, nucleus, electron shell, and the electron, proton, and neutron at the bottom of the page

Writing

- ✍ Reading Assignment: *Usborne Illustrated Dictionary of Science* pp. 126-127 (Atomic Structure)
- ✍ Additional Research Readings:
  - 📖 Atomic Structure: *USE* pg. 10-14
  - 📖 Atoms: *KSE* pp. 150-151

Dates

- ⏲ 340 BC – Aristotle proposes that all substances are made up of combinations of four elements: earth, air, water and fire.
- ⏲ 1766-1844 – John Dalton lives. He is responsible for writing the first atomic theory.
- ⏲ 1897 – J.J. Thompson discovers the electron.
- ⏲ 1909 – Rutherford, along with two other scientists, discovers the nucleus of an atom.
- ⏲ 1913 – Niels Bohr comes up with the Bohr model of an atom.

Schedules for Week 1
Two Days a Week

Day 1	Day 2
☐ Do the "Making Ice Cream" experiment ☐ Define atom, electron shell, element, and compound on SG pg. 16 ☐ Enter the dates onto the date sheets on SG pp. 8-13	☐ Read pp. 126-127 from *UIDS,* then discuss what was read ☐ Color and label the "Parts of an Atom" sketch on SG pg. 19 ☐ Prepare an outline or narrative summary, write it on SG pp. 20-21

Supplies I Need for the Week
- ✓ Heavy cream, milk, sugar, vanilla
- ✓ 1 small & 1 large zip-locking plastic bag
- ✓ Crushed ice, dish towel or oven mitt, rock salt

Things I Need to Prepare

Five Days a Week

Day 1	Day 2	Day 3	Day 4	Day 5
☐ Do the "Making Ice Cream" experiment ☐ Enter the dates onto the date sheets on SG pp. 8-13	☐ Read pp. 126-127 from *UIDS,* then discuss what was read ☐ Write an outline on SG pg. 20	☐ Define atom, electron shell, element, and compound on SG pg. 16 ☐ Color and label the "Parts of an Atom" sketch on SG pg. 19	☐ Read one or all of the additional reading assignments ☐ Write a report from what you learned on SG pg. 21	☐ Complete one of the Want More Activities listed **OR** ☐ Study a scientist from the field of Chemistry

Supplies I Need for the Week
- ✓ Heavy cream, milk, sugar, vanilla
- ✓ 1 small & 1 large zip-locking plastic bag
- ✓ Crushed ice, dish towel or oven mitt, rock salt

Things I Need to Prepare

Chemistry Unit 1: The Periodic Table ~ Week 1: Elements and Atoms

Additional Information Week 1

Notes

🕯️ **Atoms vs. Elements** – Elements are substances that are made up of one type of atom, while atoms are the smallest particles of an element that retain the chemical properties of the element. In other words an element is composed of one or more of the same type of atom. So, when you hold a lump of iron, you are holding the element iron that contains billions of iron atoms. See the following article for more information:

💻 http://elementalblogging.com/science-corner-element-and-atom/

Experiment Information

☞ **Explanation** – This experiment was meant for the students to see how much fun chemistry can be. The students will explore liquids, solids, and freezing in Unit 2, so let this just be a enjoyable experience.

☞ **Troubleshooting Tips** – Use thick plastic bags so that there is less risk of one of the bags being punctured by the ice or salt. If you can't find rock salt, you can use regular table salt instead, but be aware that it may take longer for the cream mixture to freeze.

Discussion Questions

1. What is an atom? (*UIDS pg. 126 - An atom is a microscopic particle that is made up of smaller subatomic particles called electrons, protons, and neutrons.*)
2. What is the basic structure of an atom according to modern atomic theory? (*UIDS pg. 126 - Atoms have a nucleus at the center that is composed of neutrons and protons. They have electrons that fly around the nucleus in different shells or layers.*)
3. What are the charges of the subatomic particles (i.e., protons, neutrons, and electrons)? (*UIDS pg. 126 - Protons are positively charged, electrons are negatively charged, and neutrons are neutral.*
4. What is an electron shell? (*UIDS pg. 126 - An electron shell is the region of space around the nucleus of an atom where the electrons move around.*)
5. What is an orbital? (*UIDS pg. 127 - An orbital is a region in an electron shell where one or two electrons can be found.*)
6. What is an octet, and why is it important in chemistry? (*UIDS pg. 127 - An octet is a group of eight electrons in a single electron shell. It is important because atoms with an octet in their outer electron shell are very stable and unreactive.*)
7. What is the difference between atomic number and mass number? (*UIDS pg. 127 - Atomic number tells you the number of protons in a nucleus, while mass number tells you the total number of protons and neutrons in one atom of an element.*)
8. What is an isotope? (*UIDS pg. 127 - An isotope is an atom of an element that has a different number of neutrons than another atom of the same element. Isotopes have the same atomic number, but different mass numbers.*)

Want More

↻ **Make an atom** – Have the students make a model of an atom using pom-pom balls. They can use red balls for protons (p), brown balls for neutrons (n) and yellow balls for electrons (e). Have them glue the protons and neutrons together for a nucleus. Then, glue the electrons onto yellow pipe cleaners and make a circle around the nucleus. Have the students try to make a model of helium (2p, 2n, 2e) or carbon (6p, 6n, 6e).

- **Atoms and Isotopes** – Have the students play an atoms and isotopes game. You can get directions for this game from the following blog post:
  - http://elementalscience.com/blogs/science-activities/60317571-free-chemistry-game

Sketch Week 1

Parts of an Atom

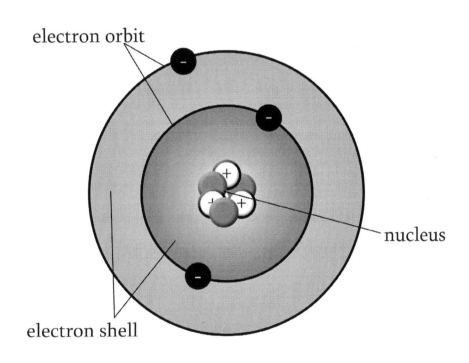

Chemistry Unit 1: The Periodic Table ~ Week 1: Elements and Atoms

Student Assignment Sheet Week 2
Periodic Table

Research Report: Element Profile Page

This week you will spend time researching an element of your choice and creating a profile page for that element. Here is what you will do:

1. Begin by choosing one of the elements of the periodic table for an in-depth profile.
2. Do some research to find out more about your chosen element. Use the internet and the resources you have in your home or at your library to find out more about the element.
3. Answer the following questions from your research:
   - ✓ What is the name, symbol, atomic number and atomic mass of the element?
   - ✓ Who discovered the element and when did that happen?
   - ✓ Is it a gas, liquid or solid at room temperature?
   - ✓ Where is the element typically found?
   - ✓ What are the major uses of the element?

   Be sure to also write down any additional interesting facts you have learned on individual index cards.
4. Finally, complete the profile page for your element. Fill out the top left-hand rectangle with the information on the element from the periodic table. Then, answer the questions and write a brief summary on the element and its uses. You may also want to include the story of how the element was discovered in your summary.

Vocabulary & Memory Work

- ☐ Vocabulary: period, group, atomic number, atomic mass
- ☐ Memory Work—This week, add the following elements to what you are working on memorizing:
   - ✓ 5-B-Boron, 6-C-Carbon, 7-N-Nitrogen, 8-O-Oxygen

Sketch: The Periodic Table

- ▨ Label the following on the periodic table: periods 1-7, groups I-VIII, color the metals yellow, the metalloids green, and the nonmetals blue, plus add a key for the color-coding
- ▨ Label the following on the element call out box: atomic number, atomic mass, symbol, element name

Writing

- ✍ Reading Assignment: *Usborne Illustrated Dictionary of Science* pp. 164-165 (The Periodic Table)
- ✍ Additional Research Readings:
   - 📖 The Periodic Table: *USE* pp. 28-29, *KSE* pp. 152-153
   - 📖 The Elements: *USE* pp. 24-25, *KSE* pp. 148-149

Dates

- ⏲ 1869 – Russian chemist Dmitri Mendeleyev draws up the very first periodic table, leaving gaps for elements that had not yet been discovered.

Schedules for Week 2

Two Days a Week

Day 1	Day 2
☐ Do the "Element Profile Page" Research Report on SG pp. 24-25 ☐ Define period, group, atomic number, and atomic mass on SG pg. 16 ☐ Enter the dates onto the date sheets on SG pp. 8-13	☐ Read pp. 164-165 from *UIDS,* then discuss what was read ☐ Color and label the "The Periodic Table" sketch on SG pg. 23 ☐ Prepare an outline or narrative summary, write it on SG pp. 26-27

Supplies I Need for the Week

Things I Need to Prepare

Five Days a Week

Day 1	Day 2	Day 3	Day 4	Day 5
☐ Do the "Element Profile Page" Research Report on SG pp. 24-25 ☐ Enter the dates onto the date sheets on SG pp. 8-13	☐ Read pp. 164-165 from *UIDS,* then discuss what was read ☐ Write an outline on SG pg. 26	☐ Define period, group, atomic number, and atomic mass on SG pg. 16 ☐ Color and label the "The Periodic Table" sketch on SG pg. 23	☐ Read one or all of the additional reading assignments ☐ Write a report from what you learned on SG pg. 27	☐ Complete one of the Want More Activities listed **OR** ☐ Study a scientist from the field of Chemistry

Supplies I Need for the Week

Things I Need to Prepare

Additional Information Week 2

Notes

- ❦ **Inner Transition Metals** – Elements 57 to 70 are also called Lanthanides, and elements 89 to 102 are also known as Actinides.

- ❦ **Superheavy Elements** – Elements that have an atomic number above 92 are known as superheavy or transuranium elements. These elements are extremely radioactive and have a very short half-life, which means that very few of these elements can actually be found in nature. Every one of these elements was first discovered in the lab. For the sake of simplicity, we will only refer to elements 104-118 as superheavy elements.

- ❦ **Discrepancies in the Periodic Table** – There are many different ways of informally grouping the elements as your students will see from the assigned and additional encyclopedia pages from this week. The periodic table can group the elements by type—metals, metalloids (or semi-metals), and nonmetals. It can group them according to orbitals—s-block, p-block, d-block, and f-block. Or it can group the elements informally—alkali metals, halogens, and so on. To learn more about why this is and download a colored versions of these tables, visit:
  💻 https://elementalscience.com/blogs/news/the-periodic-table

Experiment Information

- ☞ **Element Profile Page** – This week your students will be doing a mini-research report. They will be choosing an element from the periodic table to profile. All the instructions they need are on the Student Assignment Page, but they may need you to walk them through the process depending on how much experience they have with doing research prior to this assignment.

Discussion Questions

1. How are the elements on the periodic table organized? (*UIDS pg. 164 - The elements on the periodic table are arranged according their atomic number.*)
2. Who is credited with the basis of modern periodic table? (*UIDS pg. 164 - Dmitri Mendeleyev, a Russian chemist, is credited with the version of the periodic table that became the basis of the modern periodic table.*)
3. How does the periodic table work? (*UIDS pg. 164 - The periodic table has elements arranged horizontally in periods with increasing atomic mass. When you reach the end of a period it means that the element's outer electron shell is full.*)
4. What happens when you go across a period? (*UIDS pg. 164 - As you go across a period, the number of electrons increases by one.*)
5. What happens when you go down a group? (*UIDS pg. 165 - As you down a group on the periodic table the number of electron shells increases by one for each element.*)
6. What is the difference between metals, metalloids, and non-metals? (*UIDS pg. 165 - A metal has the metallic physical properties; these elements are solids, shiny, good conductors, and generally have high melting and boiling points. A nonmetal does not have these characteristics. Metalloids have some, but not all of these properties.*)

Want More

- ✂ **Wall Periodic Table** – Have your student make their own periodic table to display on the

wall or purchase a visual representation of the periodic table from one of the following sites:
- 💻 http://elements.wlonk.com/ (*You can also download a 8 ½ by 11 pdf version for free.*)
- 💻 http://periodictable.com/ (*This is my personal favorite, but it is pricey.*)

I have also included a blank periodic table in the Appendix on pg. 271 that you could increase in size for your students to use on the wall.

Sketch Week 2

The Periodic Table of Elements

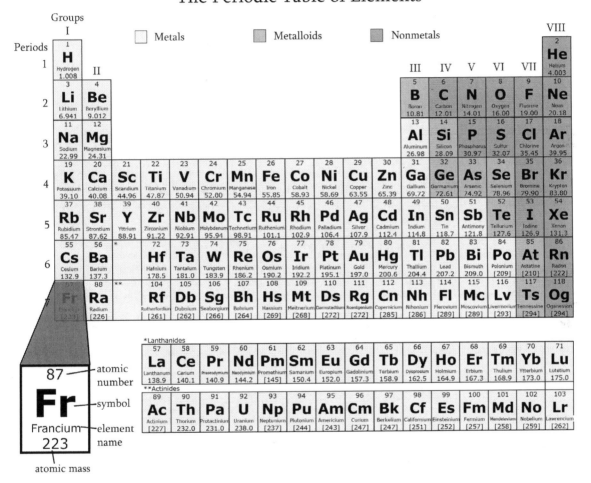

Student Assignment Sheet Week 3
Metals

Experiment: Can I transfer metal atoms?

Materials:

- ✓ Vinegar
- ✓ Salt
- ✓ 6 pennies
- ✓ Glass cup
- ✓ 2 iron nails

Procedure:

1. Read the introduction to this experiment and fill in your hypothesis.
2. Cover the bottom of your cup with a thin layer of salt. Add the pennies and cover them with vinegar.
3. After 10 minutes, take the pennies out and set them on a paper towel. Be sure to reserve the vinegar mixture.
4. Add one of the nails to the vinegar solution and leave the other one next to the glass. Let both of them sit undisturbed for 45 minutes. While you are waiting, observe what is happening to the pennies and write down what you see on your experiment sheet.
5. Take the nail out of the vinegar solution and compare it to the nail that was left outside the glass. Write your observations down on your experiment sheet.
6. Draw conclusions and complete your experiment sheet.

Vocabulary & Memory Work

- ☐ Vocabulary: metal
- ☐ Memory Work—This week, add the following elements to what you are working on memorizing:
  - ✓ 9-F-Fluorine, 10-Ne-Neon, 11-Na-Sodium, 12-Mg-Magnesium

Sketch: Metals of the Periodic Table

🖾 Label the following: the atomic number and symbol for the alkali metals, the alkaline earth metals and the transition metals. Then color the alkali metals red, the alkaline earth metals orange and the transition metals yellow.

Writing

- ᧞ Reading Assignment: *Usborne Illustrated Dictionary of Science* pg. 168 (Group I, Alkali Metals), pg. 170 (Group II, Alkaline Earth Metals), and pp. 172-173 (Transition Metals)
- ᧞ Additional Research Readings:
  - 📖 Metals: *KSE* pg. 183, Copper *KSE* pg. 199
  - 📖 Metals: *USE* pp. 30-31

Dates

- 🕐 4000 BC – Metals like lead and silver are melted down from ores and used.

Schedules for Week 3
Two Days a Week

Day 1	Day 2
☐ Do the "Can I transfer metal atoms?" experiment, then fill out the experiment sheet on SG pg. 30-31 ☐ Define metal on SG pg. 16 ☐ Enter the dates onto the date sheets on SG pp. 8-13	☐ Read pp. 168,170, 172-173 from *UIDS,* then discuss what was read ☐ Label the "Metals of the Periodic Table" sketch on SG pg. 29 ☐ Prepare an outline or narrative summary, write it on SG pg. 32-33

Supplies I Need for the Week
- ✓ Vinegar, salt
- ✓ 6 pennies
- ✓ Glass cup, 2 iron nails

Things I Need to Prepare

Five Days a Week

Day 1	Day 2	Day 3	Day 4	Day 5
☐ Do the "Can I transfer metal atoms?" experiment, then fill out the experiment sheet on SG pg. 30-31 ☐ Enter the dates onto the date sheets on SG pp. 8-13	☐ Read pp. 168,170, 172-173 from *UIDS,* then discuss what was read ☐ Write an outline on SG pg. 32	☐ Define metal on SG pg. 16 ☐ Label the "Metals of the Periodic Table" sketch on SG pg. 29	☐ Read one or all of the additional reading assignments ☐ Write a report from what you learned on SG pg. 33	☐ Complete one of the Want More Activities listed **OR** ☐ Study a scientist from the field of Chemistry

Supplies I Need for the Week
- ✓ Vinegar, salt
- ✓ 6 pennies
- ✓ Glass cup, 2 iron nails

Things I Need to Prepare

Additional Information Week 3

Experiment Information

☞ **Introduction** – (*from the Student Guide*) Copper is a metal that has many uses. It is found in the Earth's crust in a pure state. This metal is an excellent conductor, which is why it is used in wiring. It is also used in roofing, coins and decorative statues. Over prolonged exposure to the air, it will develop a dull green finish, like the Statue of Liberty. In this experiment, you are going to see if you can transfer copper atoms to an iron nail.

☞ **Results** – The students should see that over time the pennies develop a dull bluish tint after being removed from the vinegar solution. They should also see that the iron nail from the vinegar solution has a thin, shiny brownish coating on it. The coating may or may not be completely covering the nail.

☞ **Explanation** – As the pennies sat in the vinegar solution, the acid in the vinegar stripped a micro-thin layer of copper and impurities from the coins. When you first remove the pennies, they appear cleaner, but over time they developed a bluish tint. This color is the beginning of oxidation, the process which produces the dull green finish on copper. The time spent in the vinegar and salt solution sped up the process of the copper combining with the oxygen in the air to form a copper oxide compound, which has a blue-green hue. After the pennies were removed, the vinegar solution contained copper ions. Over time, they adhered to the iron nail, giving it a thin, shiny brownish coating of copper. This experiment gives the student a glimpse of how copper plating works.

☞ **Take it Further** – Pennies that were made before 1982 have a greater copper content. Repeat the experiment using pennies that are made before 1982 to see if the coating on the iron nail is thicker and more complete. (*The student should see that the nail had a thicker and more complete coating of copper because there were more copper ions available in the solution.*)

Discussion Questions

1. What are some of the characteristics of alkali metals? (*UIDS pg. 168 - Alkali metals are metals that react with water to form alkaline solutions. They are typically soft, silver-white metals. The further down the group, the more reactive the alkali metal elements are.*)

2. What are some of the characteristics of alkaline earth metals? (*UIDS pg. 170 - Alkaline earth metals are reactive metals. With the exception of Beryllium, which is a hard, white metal, the alkaline earth metals are soft, silver-white metals. The further down the group, the more reactive the alkaline earth metal elements are.*)

3. What are some of the characteristics of transition metals? (*UIDS pg. 172-173 - Transition metals are all hard, tough, shiny, and malleable metals. These metals also typically conduct heat and electricity. The transition metals also tend to have high melting points, boiling points and densities.*)

4. Choose one of the metal elements and share a few facts you have learned about that element. (*Answers will vary.*)

Want More

✐ **Make Element Trading Cards** – Have your students make an index card for several or all of the metallic elements. Each element card should include the element's name, abbreviation, atomic number and mass. The card should also say who discovered the element, when they discovered it and some common uses of the element. They can choose to write out the uses or illustrate them. See the Appendix pg. 254 for a template to use for the project. (*Note—You can use the completed set to play games in week 5.*)

- **Alkali Metals Video** – The following video is a good visual demonstration of what happens when alkali metals are mixed with water:

  🖥 http://www.youtube.com/watch?feature=player_embedded&v=m55kgyApYrY

Sketch Week 3

I have included the elements' name and mass on the key for your knowledge. The students are only required to write the symbol and atomic number for their sketch.

1 **H** Hydrogen 1.008																		
3 **Li** Lithium 6.941	**4** **Be** Beryllium 9.012																	
11 **Na** Sodium 22.99	**12** **Mg** Magnesium 24.31																	
19 **K** Potassium 39.10	**20** **Ca** Calcium 40.08	**21** **Sc** Scandium 44.96	**22** **Ti** Titanium 47.87	**23** **V** Vanadium 50.94	**24** **Cr** Chromium 52.00	**25** **Mn** Manganese 54.94	**26** **Fe** Iron 55.85	**27** **Co** Cobalt 58.93	**28** **Ni** Nickel 58.69	**29** **Cu** Copper 63.55	**30** **Zn** Zinc 65.39							
37 **Rb** Rubidium 85.47	**38** **Sr** Strontium 87.62	**39** **Y** Yttrium 88.91	**40** **Zr** Zirconium 91.22	**41** **Nb** Niobium 92.91	**42** **Mo** Molybdenum 95.94	**43** **Tc** Technetium 98.91	**44** **Ru** Ruthenium 101.1	**45** **Rh** Rhodium 102.9	**46** **Pd** Palladium 106.4	**47** **Ag** Silver 107.9	**48** **Cd** Cadmium 112.4							
55 **Cs** Cesium 132.9	**56** **Ba** Barium 137.3		**72** **Hf** Hafnium 178.5	**73** **Ta** Tantalum 181.0	**74** **W** Tungsten 183.9	**75** **Re** Rhenium 186.2	**76** **Os** Osmium 190.2	**77** **Ir** Iridium 192.2	**78** **Pt** Platinum 195.1	**79** **Au** Gold 197.0	**80** **Hg** Mercury 200.6							
87 **Fr** Francium [223]	**88** **Ra** Radium [226]																	

Student Assignment Sheet Week 4
The Inbetweens

Experiment: Is it a metal?

Materials:
- ✓ Magnet
- ✓ Materials for circuit (Flashlight bulb, Copper wire, D battery, Electrical tape - *see Appendix pg. 259*)
- ✓ Hammer
- ✓ Paper folded into a small square
- ✓ Metal paperclip (not plastic coated)
- ✓ Aluminum foil
- ✓ CD
- ✓ Safety glasses

Procedure:
1. Read the introduction to this experiment and fill in your hypothesis.
2. Conduct the four tests below for each of your materials. Fill in the chart on your experiment sheet as you go.
   - **Luster Test** – Hold your sample material up to the light. Observe whether or not it appears to shine. Does it reflect the light?
   - **Magnetism Test** – Hold a magnet close to your sample material. Is the sample attracted to the magnet?
   - **Conductivity Test** – Build the circuit using the directions found on pg. 259 in the appendix of this guide. Place your sample material in the path of the circuit. Does the light bulb light up?
   - **Malleability Test** – Take your sample material outside and place it on a hard, safe surface. After you put on your safety glasses, strike the material with a hammer and observe what happens. Was your sample flattened by the hammer or did it break apart?

   If you answered yes to all four tests, your material is most likely a metal. If you answered no to all the questions your sample material is most likely a nonmetal. If you answered both yes and no to the questions, your sample material is either a poor metal or a semimetal.
3. Draw conclusions and complete your experiment sheet.

Vocabulary & Memory Work
- ☐ Vocabulary: poor metal, semimetal, malleable, semiconductor
- ☐ Memory Work—This week, add the following elements to what you are working on memorizing:
  - ✓ 13-Al-Aluminum, 14-Si-Silicon, 15-P-Phosphorus, 16-S-Sulfur

Sketch: Group III to VI of the Periodic Table
- Label the following: the atomic number and symbol for the Group III, IV, V, and VI elements. Then, color the metals yellow, the metalloids green, and the nonmetals blue.

Writing
- Reading Assignment: *Usborne Illustrated Dictionary of Science* pg. 176 (Group III Elements), pg. 177 (Group IV Elements), pg. 180 (Group V Elements) and pp. 183 (Group VI Elements)
- Additional Research Readings:
  - Sulfur: *KSE* pg. 181, Aluminum: *KSE* pg. 200, Groups of Metals: *USE* pp. 32-33

Dates
- 1954 – Morris Tanenbaum invents the first silicon transistor at Bell Labs, beginning the use of this semiconductor in electronics.

Chemistry Unit 1: The Periodic Table ~ Week 4: The Inbetweens

Schedules for Week 4
Two Days a Week

Day 1	Day 2
☐ Do the "Is it a metal?" experiment, then fill out the experiment sheet on SG pp. 36-37 ☐ Define poor metal, semimetal, malleable, and semiconductor on SG pg. 17 ☐ Enter the dates onto the date sheets on SG pp. 8-13	☐ Read pp. 176, 177, 180, and 183 from *UIDS,* then discuss what was read ☐ Color and label the "Group III to VI of the Periodic Table" sketch on SG pg. 35 ☐ Prepare an outline or narrative summary, write it on SG pp. 38-39

Supplies I Need for the Week
✓ Magnet, Materials for circuit (Flashlight bulb, Copper wire, D battery, Electrical tape)
✓ Hammer, Paper folded into a small square, Metal paperclip (not plastic coated)
✓ Aluminum foil, CD, Safety glasses

Things I Need to Prepare

Five Days a Week

Day 1	Day 2	Day 3	Day 4	Day 5
☐ Do the "Is it a metal?" experiment, then fill out the experiment sheet on SG pp. 36-37 ☐ Enter the dates onto the date sheets on SG pp. 8-13	☐ Read pp. 176, 177, 180, and 183 from *UIDS,* then discuss what was read ☐ Write an outline on SG pg. 38	☐ Define poor metal, semimetal, malleable, and semiconductor on SG pg. 17 ☐ Color and label the "Group III to VI of the Periodic Table" sketch on SG pg. 35	☐ Read one or all of the additional reading assignments ☐ Write a report from what you learned on SG pg. 39	☐ Complete one of the Want More Activities listed **OR** ☐ Study a scientist from the field of Chemistry

Supplies I Need for the Week
✓ Magnet, Materials for circuit (Flashlight bulb, Copper wire, D battery, Electrical tape)
✓ Hammer, Paper folded into a small square, Metal paperclip (not plastic coated)
✓ Aluminum foil, CD, Safety glasses

Things I Need to Prepare

Additional Information Week 4

Experiment Information

☞ **Introduction** – (*from the Student Guide*) Metals are typically shiny and hard. They are also good conductors of electricity and heat. Some metals can be magnetic and some are malleable. On the other hand, nonmetals are typically brittle, dull and don't conduct electricity well. Poor metals and semimetals have one or more of these characteristics which places them somewhere in between metals and a nonmetals. In this experiment, you are going to use four tests to determine if a variety of materials are metals, nonmetals, or somewhere in between.

☞ **Results** – The students chart should look like:

Material	Luster Test	Magnetism Test	Conductivity Test	Malleability Test
Paper	No	No	No	No
Paperclip	Yes	Yes	Yes	Yes
Aluminum Foil	Yes	No	Yes	Yes
CD	Yes	No	No	No

☞ **Explanation** – Paper is composed of wood pulp, which is an organic material composed of nonmetal compounds. This explains why the paper failed all of the tests for being a metal. On the other hand, a paperclip is made from steel, which is a blend of several metals. This explains why the paperclip passed all the tests for being a metal. The foil is made from aluminum, which is a poor metal. This is why the foil passed the luster, malleability and conductivity tests, but not the magnetism test. A CD is made from poly-carbonate plastic, which is a nonmetal, with a thin coating of aluminum, which is a poor metal, on the outside. This explains why it only passed the luster test.

☞ **Take it Further** – Have the students choose several more materials from around your house to test. Repeat the four tests you did during the experiment to determine if these new materials are metals or not.

Discussion Questions

1. What are the main characteristics of the Group III elements? (*UIDS pg. 176 - The Group III elements are not as reactive as the Group I and II elements. Boron is a brown powder or as yellow crystals. Aluminum is a widely used white metal. The rest of the elements in Group III are soft, silver-white metals.*)

2. What are the main characteristics of the Group IV elements? (*UIDS pg. 177 - The Group IV elements are generally not very reactive and the elements become more metallic as you go down the group. Carbon is a nonmetal, silicon and germanium are both metalloids, and the remaining elements in the group are soft, silver-white metals.*)

3. What are the main characteristics of the Group V elements? (*UIDS pg. 180 - The Group V elements become more metallic as you go down the group. Nitrogen and phosphorus are both nonmetals, arsenic is a metalloid, and the remaining elements in the group are metals.*)

4. What are the main characteristics of the Group VI elements? (*UIDS pg. 183 - The Group VI elements become more metallic and less reactive as you go down the group. Oxygen*

and sulfur are both nonmetals, selenium and tellurium are a metalloids, and the remaining element, polonium is a radioactive metal.)

Want More

↻ **Make Element Trading Cards** – Have your students make an index card for several or all of the metallic elements. Each element card should include the element's name, abbreviation, atomic number and mass. The card should also say who discovered the element, when they discovered it and some common uses of the element. They can choose to write out the uses or illustrate them. See the Appendix pg. 254 for a template to use for the project. (***Note**—You can use the completed set to play games in week 5.*)

↻ **Research Report** – Silicone is one of the most abundant solids found on earth. It also has many different uses. Have your students research some of the most common ways to find silicone along with several of its most notable uses. Then, have them write a brief report sharing what they have learned.

Sketch Week 4

I have included the elements' name and mass on the key for your knowledge. The students are only required to write the symbol and atomic number for their sketch.

													5 B Boron 10.81	6 C Carbon 12.01	7 N Nitrogen 14.01	8 O Oxygen 16.00		
													13 Al Aluminum 26.98	14 Si Silicon 28.09	15 P Phosphorus 30.97	16 S Sulfur 32.07		
													31 Ga Gallium 69.72	32 Ge Germanium 72.61	33 As Arsenic 74.92	34 Se Selenium 78.96		
													49 In Indium 114.8	50 Sn Tin 118.7	51 Sb Antimony 121.8	52 Te Tellurium 127.6		
													81 Tl Thallium 204.4	82 Pb Lead 207.2	83 Bi Bismuth 209.0	84 Po Polonium [209]		

Student Assignment Sheet Week 5
Halogens and Noble Gases

Activity: The Periodic Table Game
  Materials:
  ✓ Element cards
  Procedure:
  1. Make an index card for all of the elements of the periodic table, if you have not already done so. Each element card should include the element's name, abbreviation, its atomic number and mass. The card should also say who discovered the element, when they discovered it, and some common uses of the element. You can choose to write out the uses or illustrate them. (***Note**—See the Appendix pg. 254 for a template to use when making the cards.*)
  2. Once you complete the set you can play Guess the Element or Periodic Table Bingo. (***Note**—See the Appendix pg. 253 for directions for the games.*)

Vocabulary & Memory Work
  ☐ Vocabulary: nonmetal
  ☐ Memory Work—This week, add the following elements to what you are working on memorizing:
  ✓ 17-Cl-Chlorine, 18-Ar-Argon, 19-K-Potassium, 20-Ca-Calcium

Sketch: Group VII and VIII of the Periodic Table
  ▨ Label the following: the atomic number and symbol for Group VII, also known as the halogens, and Group VIII, also known as the noble gases. Then, color the halogens blue and the noble gases purple.

Writing
  ✍ Reading Assignment: *Usborne Illustrated Dictionary of Science* pp. 186-188 (Group VII Elements), pg. 189 (Group VIII Elements)
  ✍ Additional Research Readings:
  📖 Noble Gases: *KSE* pg. 180
  📖 Halogens: *KSE* pg. 182
  📖 Carbon, Sulfur, and Phosphorus: *USE* pp. 50-56

Dates
  🕐 1722 – Antoine Lavoisier shows that diamonds are a form of carbon by burning samples of charcoal and diamond. He finds that neither produced any water and that both released the same amount of carbon dioxide per gram.
  🕐 1772 & 1774 – Swedish chemist Carl Scheele and English chemist Joseph Priestley both discover oxygen.

Schedules for Week 5
Two Days a Week

Day 1	Day 2
☐ Make your elemental cards and play one of the games ☐ Define nonmetal on SG pg. 17 ☐ Enter the dates onto the date sheets on SG pp. 8-13	☐ Read pp. 186-189 from *UIDS*, then discuss what was read ☐ Color and label the "Group VII and VIII of the Periodic Table" sketch on SG pg. 41 ☐ Prepare an outline or narrative summary, write it on SG pg. 42-43

Supplies I Need for the Week
✓ Element cards

Things I Need to Prepare

Five Days a Week

Day 1	Day 2	Day 3	Day 4	Day 5
☐ Make your elemental cards and play one of the games ☐ Enter the dates onto the date sheets on SG pp. 8-13	☐ Read pp. 186-189 from *UIDS*, then discuss what was read ☐ Write an outline on SG pg. 42	☐ Define nonmetal on SG pg. 17 ☐ Color and label the "Group VII and VIII of the Periodic Table" sketch on SG pg. 41	☐ Read one or all of the additional reading assignments ☐ Write a report from what you learned on SG pg. 43	☐ Complete one of the Want More Activities listed **OR** ☐ Study a scientist from the field of Chemistry ☐ Take the Unit 1 Test

Supplies I Need for the Week
✓ Element cards

Things I Need to Prepare

Chemistry Unit 1: The Periodic Table ~ Week 5: Halogens and Noble Gases

Additional Information Week 5

Notes

✦ **Experiment Information** – There is no experiment this week. Instead, your students will be playing two different games to help them better understand the periodic table and the elements. If your students don't want to make their own element cards, you can purchase a set from Amazon. Alternatively, you can play our periodic table match-up game, which you can download here:

⌨ https://elementalscience.com/blogs/news/the-periodic-table

Or you can play the periodic table battleship game from Teach Beside Me instead:

⌨ https://teachbesideme.com/periodic-table-battleship/

Discussion Questions

1. What are two characteristics common to all halogens? (*UIDS pg. 186 - Halogens are all nonmetals with seven electrons in their outer shell, meaning they can react to form ionic or covalent bonds. They are also typically strong oxidizing agents.*)

2. What happens to the reactivity of halogens as you go down the group? (*UIDS pg. 186 - As you go down the group, the reactivity of the halogens decreases.*)

3. What are some of the uses of halogens? (*UIDS pp. 186-188 - Students could include the following: Fluorine is used to make slippery, plastic coatings. Chlorine is used in sterilization products and in table salt. Bromine is used in photography, medicine, and disinfectants. Iodine is used as an antiseptic.*)

4. Why are noble gases extremely unreactive? (*UIDS pg. 189 - Noble gases are very unreactive because they have full outer electron shells.*)

5. What are several of the characteristics common to all noble gases? (*UIDS pg. 189 - Noble gases are colorless, odorless, and unreactive.*)

6. What are some of the uses of noble gases? (*UIDS pg. 189 - Helium is used to fill balloons so they float, and it is mixed with oxygen for deep sea divers. Neon is used in neon sights and fluorescent light because it glows when electricity passes through it. Xenon and argon are both used in electric light bulbs and fluorescent tubes.*)

Want More

🖰 **Halogen Reactivities** – Have the students watch the following video that shows the different reactivities of the halogens:

⌨ https://www.youtube.com/watch?v=saLvwX3_p1s

🖰 **Fluorine Test** – Fluorine is often used in toothpaste to help protect the enamel on teeth from being dissolved. Have the students test this ability using an egg. You will need two eggs, fluoride toothpaste, plastic wrap, white vinegar, and an egg. Coat one of the eggs with the toothpaste, wrap it in plastic wrap, and set it in the fridge overnight. After 24 hours, gently rinse off any excess toothpaste with warm water and mark it with an "F" using a permanent maker. Then, set both eggs in a cup, cover them with vinegar, and watch what happens. (*The students should see that the egg marked with an "F" does not dissolve nearly as quickly as the one without.*)

Sketch Week 5

I have included the elements' name and mass on the key for your knowledge. The students are only required to write the symbol and atomic number for their sketch.

																	2 **He** Helium 4.003
													9 **F** Fluorine 19.00	10 **Ne** Neon 20.18
													17 **Cl** Chlorine 35.45	18 **Ar** Argon 39.95
													35 **Br** Bromine 79.90	36 **Kr** Krypton 83.80
													53 **I** Iodine 126.9	54 **Xe** Xenon 131.3
													85 **At** Astatine [210]	86 **Rn** Radon [222]

Chemistry Unit 1: The Periodic Table ~ Week 5: Halogens and Noble Gases

Unit 1: The Periodic Table
Unit Test Answers

Vocabulary Matching

1. E
2. C
3. A
4. L
5. B

6. D
7. G
8. J
9. K
10. M

11. F
12. I
13. N
14. H

True or False

1. False (*The nucleus of an atom contains protons and neutrons.*)
2. False (*An octet is a group of eight electrons in a single electron shell. Atoms with an octet in their outer electron shell are very stable and unreactive.*)
3. True
4. True
5. False (*The further down the group, the more reactive the alkaline earth metal elements are.*)
6. True
7. False (*The Group V elements become more metallic as you go down the group.*)
8. True
9. True
10. False (*Halogens are nonmetals that are strong oxidizing agents.*)

Short Answer

1. Atoms have a nucleus at the center that is composed of neutrons and protons. Then they have electrons that fly around the nucleus in different shells or layers.
2. The elements in the periodic table are arranged according their atomic number.
3. Alkali metals are metals that react with water to form alkaline solutions. They are typically soft, silver-white metals. The further down the group, the more reactive the alkali metal element.
4. Answers should include two of the following from each category: Metals: Aluminum, Gallium, Indium, Tin, Antimony, Thallium, Lead, Bismuth, Polonium; Metalloids: Boron, Silicon, Germanium, Arsenic, Selenium, Tellurium; Nonmetals: Carbon, Nitrogen, Oxygen, Phosphorus, Sulfur
5. Halogens are all nonmetals with seven electrons in their outer shell, meaning they can react to form ionic or covalent bonds.
6. 1-H-Hydrogen, 2-He-Helium, 3-Li-Lithium, 4-Be-Beryllium, 5-B-Boron, 6-C-Carbon, 7-N-Nitrogen, 8-O-Oxygen, 9-F-Fluorine, 10-Ne-Neon, 11-Na-Sodium, 12-Mg-Magnesium, 13-Al-Aluminum, 14-Si-Silicon, 15-P-Phosphorus, 16-S-Sulfur, 17-Cl-Chlorine, 18-Ar-Argon, 19-K-Potassium, 20-Ca-Calcium

Unit 1: The Periodic Table
Unit Test

Vocabulary Matching

1. Atom ___

2. Electron Shell ___

3. Element ___

4. Compound ___

5. Period ___

6. Group ___

7. Atomic Number ___

8. Atomic Mass ___

9. Metal ___

10. Poor Metal ___

11. Semimetal ___

12. Malleable ___

13. Semiconductor ___

14. Nonmetal ___

A. A substance made up of one type of atom, which cannot be broken down by chemical reaction to form a simpler substance.

B. A set of elements that have the same number of electron shells, shown as rows in the periodic table.

C. A region around the nucleus of an atom where a specific number of electrons can exist.

D. A column of elements in the periodic table that have similar properties, electron configurations and valencies.

E. The smallest particle of an element that retains the chemical properties of the element.

F. A group of elements that have characteristics of both metals and nonmetals.

G. The number of protons in the nucleus of an atom.

H. A class of elements that typically forms negative ions; they are usually dull solids or gases.

I. A characteristic of a metal that means it is bendable and easily shaped.

J. The average mass number of the atoms in a sample of an element.

K. The largest class of elements, usually they are shiny and solid at room temperature.

L. A substance made up of two or more different elements that are chemically joined in fixed proportions.

M. A group of metals that are soft and weak.

N. A substance that only conducts electricity under certain conditions.

True or False

1. _____ The nucleus of an atom contains protons and electrons.

46

2. _____ An octet is a group of electrons that makes an atom very unstable.

3. _____ A period is a set of elements that have the same number of electron shells. It is shown as rows in the periodic table.

4. _____ A group is a column of elements in the periodic table that has one addition electron shell as you go down the column.

5. _____ The further down the group, the less reactive the alkaline earth metal elements are.

6. _____ Transition metals are all hard, tough, shiny, and malleable metals.

7. _____ The Group V elements become less metallic as you go down the group.

8. _____ Oxygen and sulfur are both nonmetals.

9. _____ Noble gases are very unreactive because they have full outer electron shells.

10. _____ Halogens are metals that are not strong oxidizing agents.

Short Answer
1. What is the basic structure of an atom?

2. How are the elements on the periodic table organized?

3. What are some of the characteristics of alkali metals?

4. From the elements in Group III to Group VI, give two elements that are considered to be metals, two elements that are considered to be metalloids, and two elements that are considered to be nonmetals.

5. What are two characteristics common to all halogens?

6. Fill in the 1st 20 elements of the periodic table.

• Phosphorus	• Nitrogen	• Silicon
• Hydrogen	• Oxygen	• Neon
• Helium	• Chlorine	• Sulfur
• Aluminum	• Calcium	• Argon
• Beryllium	• Fluorine	• Lithium
• Boron	• Sodium	• Potassium
• Carbon	• Magnesium	

1. _____

2. _____

3. _____

4. _____

5. _____

6. _____

7. _____

8. _____

9. _____

10. _____

11. _____

12. _____

13. _____

14. _____

15. _____

16. _____

17. _____

18. _____

19. _____

20. _____

Chemistry: Unit 2

Matter

Unit 2: Matter
Overview of Study

Sequence of Study

Week 6: States of Matter
Week 7: Solid Structures
Week 8: Molecular Properties
Week 9: Gas Laws

Materials by Week

Week	Materials
6	Cup, Ice Cubes, Pot, Thermometer
7	String, Wide mouthed jar, Pencil, Pipe cleaners, Water, Borax, Scissors
8	4 clear cups, Eye dropper, Table salt, Food coloring, Water
9	2 cups, Apple juice, Timer, Partner

Vocabulary for the Unit

1. **Plasma** – The fourth state of matter that exists only at very high temperatures.
2. **Sublimation** – Occurs when a substance changes directly from a solid to a gas without changing into a liquid.
3. **Freezing Point** – The temperature at which a substance turns from a liquid into a solid.
4. **Boiling Point** – The temperature at which a substance turns from a liquid into a gas.
5. **Cleavage** – The splitting of a crystal along a certain plane.
6. **Conductivity** – The measure of a substance's ability to conduct heat or electricity.
7. **Lattices** – The regular arrangement of repeating patterns of atoms, molecules or ions in a solid structure.
8. **Density** – A measure of the amount of matter in a substance compared to its volume.
9. **Elasticity** – The ability of a substance to stretch and then return to its original shape.
10. **Intermolecular forces** – The forces that exist between molecules, which explain properties like elasticity, surface tension, and viscosity.
11. **Diffusion** – The spreading out of a gas to fill the available space of the container it is in.
12. **Kinetic Theory** – The theory that states that as the temperature rises, particles move faster and therefore take up more space.
13. **Pressure** – The amount of force that pushes on a given area.
14. **Temperature** – A measure of the heat energy that a substance contains.
15. **Volume** – A measurement of the space occupied by a substance.

Memory Work for the Unit

The Elements of the Periodic Table – The following elements will be memorized in this unit:
- ✓ 21-Sc-Scandium
- ✓ 22-Ti-Titanium
- ✓ 23-V-Vanadium
- ✓ 24-Cr-Chromium
- ✓ 25-Mn-Manganese
- ✓ 26-Fe-Iron
- ✓ 27-Co-Cobalt
- ✓ 28-Ni-Nickel
- ✓ 29-Cu-Copper
- ✓ 30-Zn-Zinc
- ✓ 31-Ga-Gallium
- ✓ 32-Ge-Germanium
- ✓ 33-As-Arsenic
- ✓ 34-Se-Selenium
- ✓ 35-Br-Bromine
- ✓ 36-Kr-Krypton

Gas Laws
1. **Boyle's Law** – At constant temperature, the volume of a gas is inversely proportional to the pressure.
2. **Charles' Law** – At constant pressure, the volume of a gas is directly proportional to the temperature. This law is also known as the Law of Volumes.

Notes

Student Assignment Sheet Week 6
States of Matter

Experiment: When does water change state?

Materials:
- ✓ Cup
- ✓ Ice cubes
- ✓ Pot
- ✓ Thermometer

Procedure:
1. Read the introduction to this experiment and make a hypothesis.
2. Fill a small pot halfway with ice cubes. Place the pot on a burner and turn the burner onto medium heat. Observe the thermometer as the ice begins to melt and record the temperature once all of the ice melts.
3. Continue to heat the water, observing the temperature on the thermometer as it heats up. Once you begin to see the water boiling and observe the presence of steam, record your last temperature measurement.
4. Turn the burner off and remove the pot from the burner before you draw conclusions and complete your experiment sheet.

Vocabulary & Memory Work
- ☐ Vocabulary: plasma, freezing point, boiling point, sublimation
- ☐ Memory Work—This week, add the following elements to what you are working on memorizing:
  - ✓ 21-Sc-Scandium
  - ✓ 22-Ti-Titanium
  - ✓ 23-V-Vanadium
  - ✓ 24-Cr-Chromium

Sketch: States of Matter
- ▣ Label the following: gas, liquid, solid, freezing, melting, condensing, evaporating, sublimation, deposition

Writing
- ✍ Reading Assignment: *Usborne Illustrated Dictionary of Science* pp. 120-121 (States of Matter)
- ✍ Additional Research Readings:
  - 📖 Solids, Liquids, and Gases: *USE* pp. 16-17, Changes in State: *USE* pp. 18-19
  - 📖 States of Matter: *KSE* pp. 156-157

Dates
- 🕐 1897 – Sir William Crookes discovers a fourth state of matter through his experiments with gases.
- 🕐 1928 – US scientist Irving Langmuir names the fourth state of matter plasma.

Schedules for Week 6
Two Days a Week

Day 1	Day 2
☐ Do the "When does water change state?" experiment, then fill out the experiment sheet on SG pp. 50-51 ☐ Define plasma, freezing point, boiling point, and sublimation on SG pg. 46 ☐ Enter the dates onto the date sheets on SG pp. 8-13	☐ Read pp. 120-121 from *UIDS*, then discuss what was read ☐ Color and label the "States of Matter" sketch on SG pg. 49 ☐ Prepare an outline or narrative summary, write it on SG pp. 52-53

Supplies I Need for the Week
- ✓ Cup
- ✓ Water
- ✓ Pot
- ✓ Thermometer

Things I Need to Prepare

Five Days a Week

Day 1	Day 2	Day 3	Day 4	Day 5
☐ Do the "When does water change state?" experiment, then fill out the experiment sheet on SG pp. 50-51 ☐ Enter the dates onto the date sheets on SG pp. 8-13	☐ Read pp. 120-121 from *UIDS*, then discuss what was read ☐ Write an outline on SG pg. 52	☐ Define plasma, freezing point, boiling point, and sublimation on SG pg. 46 ☐ Color and label the "States of Matter" sketch on SG pg. 49	☐ Read one or all of the additional reading assignments ☐ Write a report from what you learned on SG pg. 53	☐ Complete one of the Want More Activities listed **OR** ☐ Study a scientist from the field of Chemistry

Supplies I Need for the Week
- ✓ Cup
- ✓ Water
- ✓ Pot
- ✓ Thermometer

Things I Need to Prepare

Additional Information Week 6

Notes

🐝 **Plasma** – There is a fourth state of matter, plasma, that is not mentioned in the encyclopedias. Although on earth, the three most common states of matter are solids, liquids, and gases, plasma is the most common state of matter in the universe. To teach your students about plasma, you can have them watch this video:

💻 https://www.youtube.com/watch?v=AVEGJZxglIg

Experiment Information

☞ **Introduction** – (*from the Student Guide*) There are three states of matter that are easily observed every day on Earth: solid, liquid and gas. Solids have tightly packed molecules with fixed shape and volume, liquids have widely spaced molecules with a fixed volume and gases have independently moving molecules with no fixed shape or volume. There is also a fourth state of matter called plasma, which is rarely seen on Earth. Plasma is created when a gas ionizes at very high temperatures. In this experiment, you are going to look at the points at which water changes state.

☞ **Results** – Your students should see that the water began to melt at $32°$ F ($0°$ C) and that it began to boil around $212°$ F ($100°$ C).

☞ **Explanation** – The melting point of water is $32°$ F ($0°$ C), which is the point at which it changes from a solid to a liquid. This is the reason that you began to see the ice melting to liquid water around that temperature. The boiling point of water is $212°$ F ($100°$ C), which is the point at which it changes from a liquid to a gas. This is the reason that you began to see steam rising from the pot once around that temperature.

☞ **Troubleshooting Tips** – If the temperature goes way up, but your water is not boiling, make sure that your thermometer is not touching the bottom of the pan. If it is, you will get a false reading. Also, your elevation can effect your results. The melting point and boiling point of water are given at 1 atmosphere of pressure, which is at sea level. If you live above or below sea level, this will affect the temperature at which your ice melts and your water boils. The effect will depend upon how much you are above or below sea level. Generally, the higher your elevation, the higher your temperatures will be.

☞ **Take it Further** – The students have already seen melting and evaporation. Cover the pot immediately after they take it off the burner to see condensation in action. (*The students should see droplets of water form on the lid.*) To demonstrate freezing, you can also have them place a cup of water in the freezer to see how long to takes to turn into ice. (*The students should see the liquid water turn into solid ice in 1 to 2 hours.*)

Discussion Questions

1. What are two ways that substances can change physical states? (*UIDS pg. 120 - Substances can change physical state when heated or cooled, or when the energy of the particles is increased or decreased.*)

2. What is a solid? (*UIDS pg. 120 - A solid is a state of matter that has definite shape and volume.*)

3. What is a liquid? (*UIDS pg. 120 - A liquid is a state of matter that has fixed volume, but no definite shape.*)

4. What is a gas? (*UIDS pg. 120 - A gas is a state of matter that has no fixed volume or shape.*)
5. What is condensation? (*UIDS pg. 121 - Condensation is when the particles in a gas cool off enough to become a liquid at room temperature and under normal pressure.*) Evaporation? (*UIDS pg. 121 - Evaporation is when the particles in a liquid heat up enough to turn into a gas.*)
6. What is melting? (*UIDS pg. 120 - Melting is when the particles in a solid are heated and change state into a liquid.*) Freezing? (*UIDS pg. 121 - Freezing is when the particles of a liquid cool down, turning the liquid into a solid.*)
7. What is sublimation? (*UIDS pg. 121 - Sublimation is when a solid turns directly into a gas, skipping the liquid phase.*)
8. What is one thing that can affect a boiling or melting point? (*UIDS pp. 120-121 - Pressure can both affect boiling and melting points.*)

Want More

⤴ **States of Matter Poster** – Have the students research the four states of matter (gas, solid, liquid, and plasma) and the names of the different phase changes. Next, have them create a poster with the title "States of Matter". Then, have them divide the poster into 4 sections and label each one with a different state of matter. In each box have them share what they have learned from their research in a visually attractive manner (e.g., including pictures). Finally, have the students present a brief report on the states of matter using their poster as a visual aid.

⤴ **Dry Ice Exploration** – Dry ice is the solid form of carbon dioxide. It's a fun chemical to play around with because it readily sublimes at atmospheric pressure. Check out the following article for activity suggestions and a coordinating worksheet.

🖳 http://elementalblogging.com/dry-ice-exploration/

Sketch Week 6

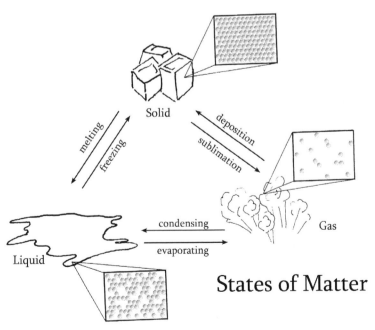

Solid

melting

freezing

deposition

sublimation

condensing

evaporating

Liquid

Gas

States of Matter

Chemistry Unit 2: Matter ~ Week 6 States of Matter

Student Assignment Sheet Week 7
Solid Structures

Experiment: Can I make crystals?

Materials:
- ✓ Wide mouthed jar
- ✓ String, Pencil, Pipe cleaner
- ✓ Water, Borax
- ✓ Scissors

Procedure:

1. Read the introduction to this experiment.
2. Fill your jar with water, pour that water into a pot and set it on a burner to boil. Make sure to record how many cups (or mL) of water it takes to fill your jar.
3. Form a shape with the pipe cleaner. This can be as simple or as complex as you wish, but make sure it will fit through the opening of your jar. Next, tie a string to the shape and then tie the other end of the string to a pencil. (*Tip—You want the pencil to be able to rest on the edge of your jar without having your shape touch the sides or bottom of the jar.*)
4. Set your jar in the sink, put on safety glasses and use a hot mitt to slowly add boiling water until it almost fills the jar. Then, add the Borax one TBSP (25 g) at a time, stirring each time until the Borax is dissolved. You want to add about 3 TBSP (75 g) of Borax for every cup (240 mL) of water you've added. (*Note—Be sure to record how many tablespoons you added on your experiment sheet.*)
5. Hang the shape in the jar so that it is completely covered by the liquid. Allow the jar to sit undisturbed overnight. In the morning, take out your shape and observe what has happened.
6. Draw conclusions and complete your experiment sheet.

> ☹ **CAUTION**
>
> **DO NOT DO THIS EXPERIMENT WITHOUT ADULT SUPERVISION!**
> Boiling hot water can cause severe damage. Be sure to use the proper safety gear.

Vocabulary & Memory Work

- ☐ Vocabulary: cleavage, conductivity, lattices
- ☐ Memory Work—This week, add the following elements to what you are working on memorizing:
  - ✓ 25-Mn-Manganese, 26-Fe-Iron, 27-Co-Cobalt, 28-Ni-Nickel

Sketch: Crystal Shapes

- ▦ Label the following: cubic, tetragonal, monoclinic, triclinic, hexagonal

Writing

- ↝ Reading Assignment: *Usborne Illustrated Dictionary of Science* pp. 135-137
- ↝ Additional Research Readings:
  - 📖 Crystals: *USE* pp. 90-91
  - 📖 Crystals: *KSE* pp. 168-169

Dates

- ⊙ 1915 – William Bragg and his son receive the Nobel Prize in Physics for their work on crystalline structures.

Schedules for Week 7
Two Days a Week

Day 1	Day 2
☐ Do the "Can I make crystals?" experiment, then fill out the experiment sheet on SG pp. 56-57 ☐ Define cleavage, conductivity, and lattices on SG pg. 46 ☐ Enter the dates onto the date sheets on SG pp. 8-13	☐ Read pp. 135-137 from *UIDS*, then discuss what was read ☐ Color and label the "Crystal Shapes" sketch on SG pg. 55 ☐ Prepare an outline or narrative summary, write it on SG pp. 58-59

Supplies I Need for the Week
✓ String, a wide mouthed jar
✓ Pencil, pipe cleaner
✓ Water, Borax
✓ Scissors

Things I Need to Prepare

Five Days a Week

Day 1	Day 2	Day 3	Day 4	Day 5
☐ Do the "Can I make crystals?" experiment, then fill out the experiment sheet on SG pp. 56-57 ☐ Enter the dates onto the date sheets on SG pp. 8-13	☐ Read pp. 135-137 from *UIDS*, then discuss what was read ☐ Write an outline on SG pg. 58	☐ Define cleavage, conductivity, and lattices on SG pg. 46 ☐ Color and label the "Crystal Shapes" sketch on SG pg. 55	☐ Read one or all of the additional reading assignments ☐ Write a report from what you learned on SG pg. 59	☐ Complete one of the Want More Activities listed **OR** ☐ Study a scientist from the field of Chemistry

Supplies I Need for the Week
✓ String, a wide mouthed jar
✓ Pencil, pipe cleaner
✓ Water, Borax
✓ Scissors

Things I Need to Prepare

Additional Information Week 7

Experiment Information

☞ **Note** – You can find Borax in the laundry aisle of the local grocery store. Be sure to buy the one labeled laundry booster, not the soap that includes Borax.

☞ **Introduction** – (*from the Student Guide*) Crystals come in all shapes and sizes, but each is based on the regularly repeating pattern of the atoms, molecules or ions that make it up. They are generally very hard solid structures and we use them in a variety of ways as we go about our daily life. For example, the salt you use on your food and the lead in your pencil are both made up of crystals. In this experiment, you are going to attempt to make your own crystals.

☞ **Results** – The students should see that their pipe cleaner creation is completely coated with crystals. The crystals should be white, making their shape look like it is covered in ice or snow.

☞ **Explanation** – You were able to super saturate the water with Borax crystals because hot water can hold more crystals than cold water. This is due to the fact that the hot water molecules are spread farther apart, which creates more room into which the Borax molecules can dissolve. As the water cooled, the molecules came closer together, causing the Borax molecules to be forced out of the solution and form solid crystals once again. The pipe cleaner gave the molecules an easy place to attach, which is why it was coated with crystals.

☞ **Take it Further**—Repeat the experiment, but this time use sugar instead of Borax. The crystals will take much longer to form (at least 1 to 2 weeks), but this time you will be able to eat them! (*The students should see the same results for their sugar crystals for the same reasons stated above.*)

Discussion Questions

1. What are crystals? (*UIDS pg. 135 - Crystals are solids with regular geometric shapes that can be formed from the regular arrangement of atoms, ions, or molecules.*)
2. What are two methods of crystallization? (*UIDS pg. 135 - Two methods of crystallization are allowing a solvent to evaporate and placing a seed crystal in a supersaturated solution.*)
3. What causes crystals to form a definite shape? (*UIDS pg. 136 - Crystals form a definite shape because of the arrangement of their atoms or ions.*)
4. What are the main crystal shapes? (*UIDS pg. 136 - The five basic shapes of crystals are cubic, tetragonal, monoclinic, triclinic [or rhombohedral], and hexagonal.*)
5. What is polymorphism in crystals? (*UIDS pg. 136 - Polymorphism in crystals is when a substance can have two or more different crystal shapes.*)
6. What are the four main type of lattice structures? Describe each one. (*UIDS pg. 137 - The four main types of lattice structures are giant atomic, giant ionic, giant metallic, and molecular lattices. Giant atomic lattices are crystal lattices held together with covalent bonding. Giant ionic lattices are crystal lattices held together with ionic bonding. Giant metallic lattices are crystal lattices held together by metallic bonding with delocalized electrons. Molecular lattices are crystal lattices held together by weak intermolecular forces.*)

Want More

- **Make Hot Ice** – This experiment is very cool when it works, but it's a bit temperamental. You will need vinegar and baking soda for the experiment. Directions and troubleshooting tips can be found at:
    - 🖥 http://chemistry.about.com/od/homeexperiments/a/make-hot-ice-sodium-acetate.htm
- **Looking at Crystals** – Have the students use a magnifying glass or microscope to examine sugar and salt crystals. You can also have them examine one of their Borax crystals from the experiment. Talk about the different crystal systems (or shapes) that the students see. (*Borax and sugar are both monoclinic, while salt is cubic.*)

Sketch Week 7

Crystal Shapes

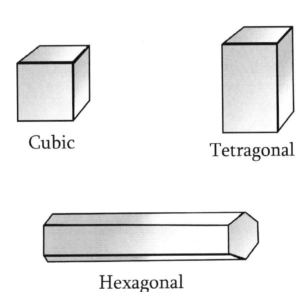

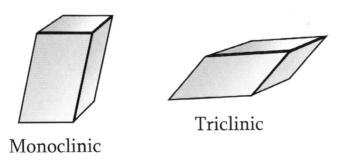

Student Assignment Sheet Week 8
Molecular Properties

Experiment: How does the addition of salt affect the density of water?

Materials:
- ✓ 4 clear cups
- ✓ Eye dropper
- ✓ Table salt
- ✓ Food coloring
- ✓ Water

Procedure:

1. Read the introduction to this experiment and make a hypothesis. Then, prepare your solutions according to the directions on the experiment sheet on SG pg. 62.
2. Label each of your cups #1-4. Fill cup #1 with 1 cup (240 mL) of the yellow solution, cup #2 with 1 cup (240 mL) of the red solution, cup #3 with 1 cup (240 mL) of the blue solution, and cup #4 with 1 cup (240 mL) of the green solution.
3. Using an eye dropper, draw up about 1-2 tsp (5-10 mL) of the yellow solution and **very slowly** add it to the cup #2. (**Note**—*When you add the yellow solution, you need to make sure the tip of your eyedropper is just below the surface of the solution in the cup.*) Wait 1 minute and record your observations.
4. Next, using an eye dropper, draw up about 1-2 tsp (5-10 mL) of the yellow solution and **very slowly** add it to the cup #3. Wait 1 minute and record your observations.
5. Then, using an eye dropper, draw up about 1-2 tsp (5-10 mL) of the yellow solution and **very slowly** add it to the cup #4. Wait 1 minute and record your observations.
6. Pour out the solutions from each of the used cups and refill them with 1 cup of each of the colors as before.
7. Repeat steps 3 through 6 with the red, blue, and green solution. (**Note**—*Each time you will skip the cup with the same color solution because we already know that they have the same concentration of salt.*)
8. Draw conclusions and complete your experiment sheet.

Vocabulary & Memory Work

- ☐ Vocabulary: density, elasticity, intermolecular forces
- ☐ Memory Work—This week, add the following elements to what you are working on memorizing:
  - ✓ 29-Cu-Copper, 30-Zn-Zinc, 31-Ga-Gallium, 32-Ge-Germanium

Sketch: Density Column

🔲 Based on what you know, label the 5 layers of the density column with these liquids: water (density 1.00 g/mL), corn syrup (density 1.38 g/mL), salt water (density 1.03 g/mL), vegetable oil (density 0.92 g/mL), and rubbing alcohol (density 0.79 g/mL)

Writing

✎ Reading Assignment: *Usborne Illustrated Dictionary of Science* pp. 22-23 (Molecular Properties), pg. 24 (Density)

✎ Additional Research Readings: How Liquids Behave: *USE* pp. 20-21

Dates

🕐 287 BC-212 BC – Archimedes lived. He used the principle of density to determine if King Hiero II's crown was made of real gold.

Chemistry Unit 2: Matter ~ Week 8 Molecular Properties

Schedules for Week 8
Two Days a Week

Day 1	Day 2
☐ Do the "How does the addition of salt affect the density of water?" experiment, then fill out the experiment sheet on SG pp. 62-63 ☐ Define density, elasticity, and intermolecular forces on SG pp. 46-47 ☐ Enter the dates onto the date sheets on SG pp. 8-13	☐ Read pp. 22-24 from *UIDS*, then discuss what was read ☐ Color and label the "Density Column" sketch on SG pg. 61 ☐ Prepare an outline or narrative summary, write it on SG pp. 64-65

Supplies I Need for the Week
- ✓ 4 clear cups
- ✓ Eye dropper
- ✓ Table salt
- ✓ Food coloring
- ✓ Water

Things I Need to Prepare

Five Days a Week

Day 1	Day 2	Day 3	Day 4	Day 5
☐ Do the "How does the addition of salt affect the density of water?" experiment, then fill out the experiment sheet on SG pp. 62-63	☐ Read pp. 22-24 from *UIDS*, then discuss what was read ☐ Write an outline on SG pg. 64	☐ Define density, elasticity, and intermolecular forces on SG pp. 46-47 ☐ Color and label the "Density Column" sketch on SG pg. 61	☐ Enter the dates onto the date sheets on SG pp. 8-13 ☐ Read the additional reading assignments ☐ Write a report from what you learned on SG pg. 65	☐ Complete one of the Want More Activities listed **OR** ☐ Study a scientist from the field of Chemistry

Supplies I Need for the Week
- ✓ 4 clear cups
- ✓ Eye dropper
- ✓ Table salt, food coloring and water

Things I Need to Prepare

64

Additional Information Week 8

Experiment Information

☞ **Introduction** – (*from the Student Guide*) Density is a property of matter that relates mass and volume. Mass tells you how much matter is in something, while volume tells you how much space something occupies. So in other words, density measures how much matter of a particular substance will occupy a given space. When comparing two different solutions, a more dense solution will sink to the bottom, while a less dense solution will float to the top. In this week's experiment, you are going to use the principle of density to see how salt affects the density water.

☞ **Results** – A student's results chart should look like:

	Added to the Yellow Solution	Added to the Red Solution	Added to the Blue Solution	Added to the Green Solution
Yellow Solution		Floated	Floated	Floated
Red Solution	Sank		Floated	Floated
Blue Solution	Sank	Sank		Floated
Green Solution	Sank	Sank	Sank	

☞ **Explanation** – The addition of salt to the water increases the overall mass of the solution, but minimally affects the volume. This means that the density will increase. The more salt you add to the water, the more dense the solution will become. This is the reason why you saw that the yellow liquid always floated to the top of each cup, while the green liquid always sank to the bottom.

☞ **Take if Further** – Have the students make a density column. You will need ½ cup (120 mL) of honey, corn syrup, dish soap, water, vegetable oil, rubbing alcohol, and lamp oil each in separate cups. Add a different shade of food coloring to each of the liquids, except the honey and vegetable oil. (*Note—If your dish soap is already colored, you do not need to add a color to that.*) Next, pour the honey into a cylindrical container. Then, slowly add the corn syrup on top. You want to pour the liquid slowly down the side of the container to minimize the bubbles and give you the best results. Follow with the dish soap, water, vegetable oil, alcohol, and finally the lamp oil. You should have seven separate liquid layers on top of each other. Each layer has the same volume, but different densities.

Discussion Questions

1. How is elasticity related to intermolecular forces? (*UIDS pg. 22 - When a material is stretched or compressed, the molecules in it are attracted or repelled to each other. So when the stretch or compression is removed from the material, the intermolecular forces cause it to return to the way it was.*)
2. What does Hooke's Law say? (*UIDS pg. 22 - Hooke's Law says that when a distorting force is applied to an object, the strain is proportional to the stress.*)
3. What is surface tension? (*UIDS pg. 23 - Surface tension is the skin-like properties of a liquid's surface due to intermolecular forces.*) Adhesion? (*UIDS pg. 23 - Adhesion is the intermolecular force of attraction between molecules within different substances.*) Cohesion?

(*UIDS pg. 23 - Cohesion is the intermolecular force of attraction between molecules within the same substance.*)

4. What is specific gravity? (*UIDS pg. 24 - Specific gravity, also known as relative density, is the density of a substance relative to the density of water.*)

5. What does a hydrometer measure? (*UIDS pg. 24 - A hydrometer measure the density of a liquid.*)

Want More

⌕ **Calculate the Density** – If the students are ready for some of the math of chemistry, have them calculate the densities of the solutions that they made in the experiment. You can use the Density Calculations worksheet on pp. 257-258 of the Appendix with your students.

Answers:

☑ Yellow solution = 1.00 g/mL   ☑ Blue solution = 1.29 g/mL

☑ Red solution = 1.14 g/mL   ☑ Green solution = 1.43 g/mL

⌕ **Surface Tension** – Have the students play with surface tension. You will need a shallow bowl, a needle, and dish soap. Fill the bowl about halfway with water. Gently place the needle on the surface without touching the water. Observe what happens to the surrounding water. (*If you look closely, the student should see that the needle looks like it is resting on the "skin" of the water formed by surface tension.*) Add a drop of dish soap near the needle and observe what happens to the needle. (*The needle should sink to the bottom because the soap molecules break the week intermolecular forces that form the surface tension "skin" of the water.*)

Sketch Week 8

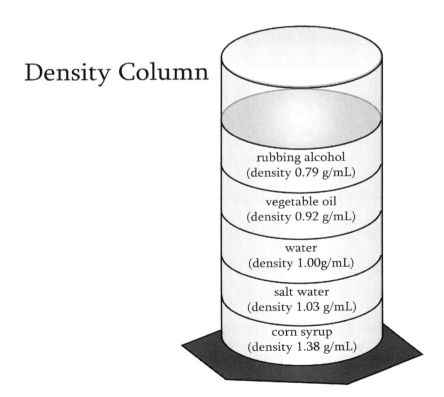

Density Column

rubbing alcohol
(density 0.79 g/mL)

vegetable oil
(density 0.92 g/mL)

water
(density 1.00g/mL)

salt water
(density 1.03 g/mL)

corn syrup
(density 1.38 g/mL)

Chemistry Unit 2: Matter ~ Week 8 Molecular Properties

Student Assignment Sheet Week 9
Gas Laws

Experiment: Does temperature affect the rate of diffusion?

Materials:
- ✓ 2 cups, apple juice, timer, partner

Procedure:
1. Read the introduction to this experiment and make a hypothesis.
2. Fill 2 cups with apple juice, cover each of them with plastic wrap, and leave them on the counter.
3. After one hour, take one of the cups from the counter and have your partner stand in the opposite corner of the room. Remove the plastic wrap from the cup, time how long it takes for them to smell the apple juice. Record that time on your experiment sheet. (*Note—If your partner doesn't smell it after 10 minutes, end the test and note the results on the experiment sheet.*)
4. Next, take the other cup from the counter and microwave it on high for 1-2 minutes, until the juice is warm, but not too hot to touch. Have your partner stand in the opposite corner of the room and remove the plastic wrap from the cup. Time how long it takes for them to smell the apple juice and record that time on your experiment sheet. (*Note—If your partner doesn't smell it after 10 minutes, end the test and note the results on the experiment sheet.*)
5. Draw conclusions and complete your experiment sheet.

> ☣ **CAUTION**
>
> **DO NOT DO THIS EXPERIMENT WITHOUT ADULT SUPERVISION!**
> The apple juice could be very hot when removing it from the microwave. Make sure to use protective gear.

Vocabulary & Memory Work
- ☐ Vocabulary: diffusion, kinetic theory, pressure, temperature, volume
- ☐ Memory Work—This week, add the following elements to what you are working on memorizing:
  - ✓ 33-As-Arsenic, 34-Se-Selenium, 35-Br-Bromine, 36-Kr-Krypton
- ☐ Memory Work—Work on memorizing Boyle's Law and Charles' Law. (*See pg. 53 for these gas laws.*)

Sketch: Gas Laws
- ▦ Label the Following – Boyle's Law: volume decrease, pressure increase, constant temperature; Charles' Law (*also knows as the Law of Volumes*): temperature increase, volume increase, constant pressure.

Writing
- ✍ Reading Assignment: *Usborne Illustrated Dictionary of Science* pp. 142-143 (Gas Laws)
- ✍ Additional Research Readings:
  - 📖 Pressure: *KSE* pg. 311, How Gasses Behave: *USE* pp. 22-23
  - 📖 Kinetic theory: *UIDS* pg. 123

Dates
- 🕐 1827 – Robert Brown observes pollen grains randomly bouncing round in water, discovering Brownian Motions, which happen because the grains are being constantly hit by water molecules.
- 🕐 1860 – Ludwig Boltzmann develops the kinetic theory of gases, which is widely opposed by other scientists.

Schedules for Week 9
Two Days a Week

Day 1	Day 2
☐ Do the "Does temperature affect the rate of diffusion?" experiment, then fill out the experiment sheet on SG pp. 68-69 ☐ Define diffusion, kinetic theory, pressure, temperature, and volume on SG pg. 47 ☐ Enter the dates onto the date sheets on SG pp. 8-13	☐ Read pp. 142-143 from *UIDS,* then discuss what was read ☐ Color and label the "Gas Laws" sketch on SG pg. 67 ☐ Prepare an outline or narrative summary, write it on SG pp. 70-71

Supplies I Need for the Week
- ✓ 2 cups
- ✓ Apple juice
- ✓ Timer
- ✓ Partner

Things I Need to Prepare

Five Days a Week

Day 1	Day 2	Day 3	Day 4	Day 5
☐ Do the "Does temperature affect the rate of diffusion?" experiment, then fill out the experiment sheet on SG pp. 68-69 ☐ Enter the dates onto the date sheets on SG pp. 8-13	☐ Read pp. 142-143 from *UIDS,* then discuss what was read ☐ Write an outline on SG pg. 70	☐ Define diffusion, kinetic theory, pressure, temperature, and volume on SG pg. 47 ☐ Color and label the "Gas Laws" sketch on SG pg. 67	☐ Read one or all of the additional reading assignments ☐ Write a report from what you learned on SG pg. 71	☐ Complete one of the Want More Activities listed **OR** ☐ Take the Unit 2 Test

Supplies I Need for the Week
- ✓ 2 cups
- ✓ Apple juice
- ✓ Timer
- ✓ Partner

Things I Need to Prepare

Additional Information Week 9

Experiment Information

☞ **Introduction** – (*from the Student Guide*) The particles in a substance are constantly in motion. In a solid, these particles are closely packed together, so they can only move by vibrating or shaking. In a liquid, these particles can move more freely, but they still remain relatively close. In a gas, these particles are widely spaced and can move very quickly. When the particles in a substance move from one corner of a room to the other, they are said to be diffusing, or spreading out. In this week's experiment, you are going to examine how temperature affects the rate of diffusion.

☞ **Results** – The student should see that the hot apple juice diffuses through the room much quicker than the room temperature apple juice. In fact, they may see that their partner never smells the room temperature apple juice.

☞ **Explanation** – As the particles in the juice heat up, their energy of motion increases. This means that they are able to escape from the glass and travel around the room quicker than the room temperature molecules. This is known as the kinetic theory, which states that as the temperature rises, the particles move faster, and therefore they take up more space.

☞ **Take it Further** – Have the student test to see if the same is true for diffusion of liquids. Have them add several drops of food coloring to a glass of cold water and time how long it takes for the color to spread out and become uniform throughout. Then, repeat with a glass of hot water. (*The student should find that the color diffuses through the hot water much quicker than the cold water for the same reasons explained above.*)

Discussion Questions

1. What are the three gas laws? (*UIDS pg. 142 - Boyle's Law says that at constant temperature, the volume of a gas is inversely proportional to the pressure. The Law of Volumes says that at constant pressure, the volume of a gas is directly proportional to the temperature. The Third Gas Law says that at constant volume, the pressure is directly proportional to the temperature.*)

2. What is an ideal gas? (*UIDS pg. 142 - An ideal gas is one that behaves in an ideal way, its molecules have no volume or intermolecular forces and move in straight lines, losing no energy when they collide.*)

3. What is Dalton's law of partial pressure? (*UIDS pg. 143 - Dalton's law of partial pressure says that the sum of the partial pressure of each gas in a mixture is equal to the total pressure.*)

4. What does Graham's law of diffusion tell us? (*UIDS pg. 143 - Graham's law of diffusion tell us that the faster a gas diffuses, the lower its density, and vice versa.*)

5. What is stp? (*UIDS pg. 143 - Stp is an abbreviation for standard temperature and pressure.*)

6. What is Avogadro's Law? (*UIDS pg. 143 - Avogadro's law states that equal volumes of gases at the same temperature and pressure contain the same number of molecules.*)

Want More

✎ **Testing Boyle's Law** – You will need a marshmallow and a large needleless syringe. Begin

Chemistry Unit 2: Matter ~ Week 9 Gas Laws

by removing the plunger and putting the marshmallow into the syringe. Replace the plunger and push it down until it is almost touching the marshmallow. Then, cover the tip of the syringe with your finger to create a vacuum and pull the plunger up. (*You should see that the marshmallow expands because as the volume increases, the pressure on the marshmallow decreases.*)

- **Testing Charles's Law** – You will need 2 balloons, water and ice. Blow each of the balloons up to the same size. Place one on the counter and one in a bucket of ice water. Wait for 10 minutes and take the balloon out of the ice water. Compare the two balloons side by side. (*You should see that the one that was in the ice water has decreased in size because as the temperature decreases, the volume of a gas will also decrease.*)

Sketch Week 9

Gas Laws

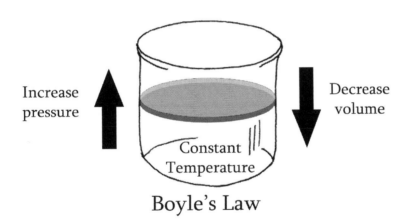

Increase pressure

Decrease volume

Constant Temperature

Boyle's Law

Charles's Law

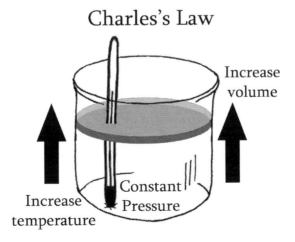

Increase volume

Constant Pressure

Increase temperature

Unit 2: Matter
Unit Test Answers

Vocabulary Matching

1. E
2. I
3. B
4. D
5. N

6. F
7. K
8. A
9. C
10. O

11. M
12. H
13. G
14. J
15. L

True or False

1. True
2. False (*The boiling point is the point at which a liquid changes into a gas and the melting point is the point at which a solid changes into a liquid.*)
3. True
4. True
5. False (*Adhesion is the intermolecular force of attraction between molecules within different substances, while cohesion is the intermolecular force of attraction between molecules within the same substance.*)
6. True
7. True
8. False (*Stp is an abbreviation for standard temperature and pressure.*)

Short Answer

1. The three states of matter that are typically found on Earth are solid, liquid and gas. A solid is a state of matter that has definite shape and volume. A liquid is a state of matter that has fixed volume, but no definite shape. A gas is a state of matter that has no fixed volume or shape.
2. Crystals form a definite shape because of the arrangement of their atoms or ions.
3. Surface tension is the skin-like properties of a liquid's surface due to intermolecular forces.
4. At constant temperature, the volume of a gas is inversely proportional to the pressure.
5. At constant pressure, the volume of a gas is proportional to the temperature.
6. 21-Sc-Scandium, 22-Ti-Titanium, 23-V-Vanadium, 24-Cr-Chromium, 25-Mn-Manganese, 26-Fe-Iron, 27-Co-Cobalt, 28-Ni-Nickel, 29-Cu-Copper, 30-Zn-Zinc, 31-Ga-Gallium, 32-Ge-Germanium, 33-As-Arsenic, 34-Se-Selenium, 35-Br-Bromine, 36-Kr-Krypton

Unit 2: Matter
Unit Test

Vocabulary Matching

1. Plasma ___

2. Sublimation ___

3. Freezing Point ___

4. Boiling Point ___

5. Cleavage ___

6. Conductivity ___

7. Lattices ___

8. Density ___

9. Elasticity ___

10. Intermolecular forces ___

11. Diffusion ___

12. Kinetic Theory ___

13. Pressure ___

14. Temperature ___

15. Volume ___

A. A measure of the amount of matter in a substance compared to its volume.

B. The temperature at which a substance turns from a liquid into a solid.

C. The ability of a substance to stretch and then return to its original shape.

D. The temperature at which a substance turns from a liquid into a gas.

E. The fourth state of matter that exists only at very high temperatures.

F. The measure of a substance's ability to conduct heat or electricity.

G. The amount of force that pushes on a given area.

H. The theory that states that as the temperature rises, particles move faster and therefore take up more space.

I. Occurs when a substance changes directly from a solid to a gas without changing into a liquid.

J. A measure of the heat energy that a substance contains.

K. The regular arrangement of repeating patterns of atoms, molecules or ions in a solid structure.

L. A measure of the space occupied by a substance.

M. The spreading out of a gas to fill the available space of the container it is in.

N. The splitting of a crystal along a certain plane.

O. The forces that exist between molecules, which explain properties like elasticity, surface tension, and viscosity.

True or False

1. _____ Substances can change physical state when heated or cooled, or when the energy of the particles is increased or decreased.

Chemistry Unit 2: Matter ~ Unit Test

2. _____ The melting point is the point at which a liquid changes into a gas and the boiling point is the point at which a solid changes into a liquid.

3. _____ Two methods of crystallization are allowing a solvent to evaporate and placing a seed crystal in a supersaturated solution.

4. _____ The four main types of lattice structures are giant atomic, giant ionic, giant metallic, and molecular lattices.

5. _____ Cohesion is the intermolecular force of attraction between molecules within different substances, while adhesion is the intermolecular force of attraction between molecules within the same substance.

6. _____ Avogadro's law states that equal volumes of gases at the same temperature and pressure contain the same number of molecules.

7. _____ An ideal gas behaves in an ideal way.

8. _____ Stp stands for super-trans-phosphorus.

Short Answer

1. What are the three states of matter typically found on Earth? (Be sure to give a characteristic of each.)

2. What causes crystals to form a definite shape?

3. What is surface tension?

4. What is Boyle's Gas Law?

5. What is Charles' Gas Law?

6. Fill in elements 21-36 from the periodic table.

• Chromium	• Iron	• Krypton
• Manganese	• Nickel	• Arsenic
• Scandium	• Selenium	• Gallium
• Vanadium	• Cobalt	• Bromine
• Titanium	• Copper	
• Germanium	• Zinc	

21. _____

22. _____

23. _____

24. _____

25. _____

26. _____

27. _____

28. _____

29. _____

30. _____

31. _____

32. _____

33. _____

34. _____

35. _____

36. _____

Chemistry: Unit 3

Solutions

Unit 3: Solutions
Overview of Study

Sequence of Study

Week 10: Compounds and Mixtures
Week 11: Solutions
Week 12: Separating Mixtures
Week 13: Electrolysis

Materials by Week

Week	Materials
10	Bag of multi-colored marshmallows, Toothpicks
11	5 clear cups (or beakers), 5 plastic spoons, Sugar, Salt, Baking powder, Flour, Cornstarch, Water, Vegetable oil, Tablespoon
12	Coffee filters, Markers, Alcohol, Coffee can or wide-mouthed jar, Rubber bands, Eyedropper
13	Distilled water, 2 test tubes, Salt, Glass cup, 2 Alligator clips, Covered copper wire, 6-volt Lantern battery, Permanent marker

Vocabulary for the Unit

1. **Molecule** – A substance that is formed when two or more atoms chemically join together.
2. **Mixture** – A combination of two or more elements or compounds that are not chemically combined.
3. **Miscible** – Liquids that can be blended together.
4. **Immiscible** – Liquids that cannot be blended together.
5. **Solution** – A homogenous mixture of two or more substances.
6. **Solubility** – The ability of a solute to be dissolved.
7. **Solute** – The substance that dissolves in the solvent to form a solution.
8. **Solvent** – The substance in which the solute dissolves to form a solution, typically a liquid.
9. **Chromatography** – A method of separating the substances in a mixture by the rate they move through or along a medium, such as filter paper.
10. **Anode** – The positively charged electrode by which current enters the cell.
11. **Cathode** – The negatively charged electrode by which current leaves the cell.
12. **Electrolyte** – A substance that conducts electricity when it is in solution.

Memory Work for the Unit

The Elements of the Periodic Table – The following elements will be memorized in this unit:
- ✓ 37-Rb-Rubidium
- ✓ 38-Sr-Strontium
- ✓ 39-Y-Yttrium
- ✓ 40-Zr-Zirconium
- ✓ 41-Nb-Niobium
- ✓ 42-Mo-Molybdenum
- ✓ 43-Tc-Technetium
- ✓ 44-Ru-Ruthenium
- ✓ 45-Rh-Rhodium
- ✓ 46-Pd-Palladium
- ✓ 47-Ag-Silver
- ✓ 48-Cd-Cadmium
- ✓ 49-In-Indium
- ✓ 50-Sn-Tin
- ✓ 51-Sb-Antimony
- ✓ 52-Te-Tellurium

Law of Constant Composition – A pure compound always contains the same elements in the same proportions.

Notes

Student Assignment Sheet Week 10
Compounds and Mixtures

Experiment: Marshmallow Molecules
   Materials:
   - ✓ Bag of multi-colored marshmallows
   - ✓ Toothpicks

   Procedure:
   1. Read the introduction to this experiment.
   2. Choose a color of marshmallow to represent each of the following atoms: oxygen, nitrogen, hydrogen and carbon. When you make your molecules you must follow these rules:
      → Oxygen prefers to bond twice, nitrogen prefers to bond three times, hydrogen prefers to only bond once and carbon prefers to bond 4 times;
      → Each atom must have its preferred number of bonds to form a stable molecule and you must only create stable molecules;
      → All of nitrogen's bonds should point down and all of carbon's bonds need to be opposite from each other (except in the case of a multiple bond).
   3. Make the following molecules from your marshmallows: NH_3 (ammonia), H_2O (water), CH_4 (methane), CO_2 (carbon dioxide), C_2H_5OH (ethanol) (**Hint**—*You can make multiple bonds between the atoms.*)
   4. Draw a quick sketch of each of your molecules after you finish assembling them.
   5. Draw conclusions and complete your experiment sheet.

Vocabulary & Memory Work
   - ☐ Vocabulary: molecule, mixture, miscible, immiscible
   - ☐ Memory Work—This week, add the following elements to what you are working on memorizing:
      - ✓ 37-Rb-Rubidium, 38-Sr-Strontium, 39-Y-Yttrium, 40-Zr-Zirconium
   - ☐ Memory Work—Work on memorizing the Law of Constant Composition: A pure compound always contains the same elements in the same proportions.

Sketch—No sketch this week.

Writing
   - ✍ Reading Assignment: *Usborne Illustrated Dictionary of Science* pp. 122-123 (Elements, Compounds, and Mixtures - except the section on kinetic theory)
   - ✍ Additional Research Readings:
      - 📖 Chemical Compounds: *KSE* pp. 164-165
      - 📖 Molecules: *USE* pp. 14-15

Dates
   - ⏲ 1649 – Pierre Gassendi states that atoms can be joined together to form molecules.
   - ⏲ 1798-1808 – Joseph-Louis Proust analyzes the different sources of several compounds and finds that their elements always contained the same ratio by weight. This leads to the discovery of the law of constant composition.

Chemistry Unit 3: Solutions ~ Week 10 Compounds and Mixtures

Schedules for Week 10
Two Days a Week

Day 1	Day 2
☐ Do the "Marshmallow Molecules" experiment, then fill out the experiment sheet on SG pp. 77-79 ☐ Define molecule, mixture, miscible, and immiscible on SG pg. 74 ☐ Enter the dates onto the date sheets on SG pp. 8-13	☐ Read pp. 122-123 from *UIDS,* then discuss what was read ☐ Prepare an outline or narrative summary, write it on SG pp. 80-81

Supplies I Need for the Week
✓ Bag of multi-colored marshmallows
✓ Toothpicks

Things I Need to Prepare

Five Days a Week

Day 1	Day 2	Day 3	Day 4	Day 5
☐ Do the "Marshmallow Molecules?" experiment, then fill out the experiment sheet on SG pp. 77-79	☐ Read pp. 122-123 from *UIDS,* then discuss what was read ☐ Write an outline on SG pg. 80	☐ Define molecule, mixture, miscible, and immiscible on SG pg. 74 ☐ Enter the dates onto the date sheets on SG pp. 8-13	☐ Read one or all of the additional reading assignments ☐ Write a report from what you learned on SG pg. 81	☐ Complete one of the Want More Activities listed **OR** ☐ Study a scientist from the field of Chemistry

Supplies I Need for the Week
✓ Bag of multi-colored marshmallows
✓ Toothpicks

Things I Need to Prepare

80

Additional Information Week 10

Notes

❧ **Molecules vs. Compounds** – Molecules are formed when two or more atoms join together. Compounds are formed when two or more elements join together. For example H_2 (*hydrogen gas*) is a molecule because two atoms of hydrogen are joined together. However, since there is only one type of element present, H_2 is not a compound. On the other hand, H_2O (*water*) is a molecule because the three atoms, one oxygen atom and two hydrogen atoms, have been joined together to form it. It is also a compound because it contains two different elements, hydrogen and oxygen. So, all compounds are molecules, but not all molecules are compounds.

Experiment Information

☞ **Introduction** – (*from the Student Guide*) Molecules are formed when two or more atoms are held together by a chemical bond. These molecules form the millions of substances that surround us in daily life, such as water and table salt. The atoms tend to follow a set of rules when they form molecules. In this experiment, you are going to make marshmallow representations of several common molecules.

☞ **Results** – This experiment is meant to be a demonstration of how molecules are formed so that the concepts will be solidified in your students' mind. The pictures below will give you an idea of what their creations should look like:

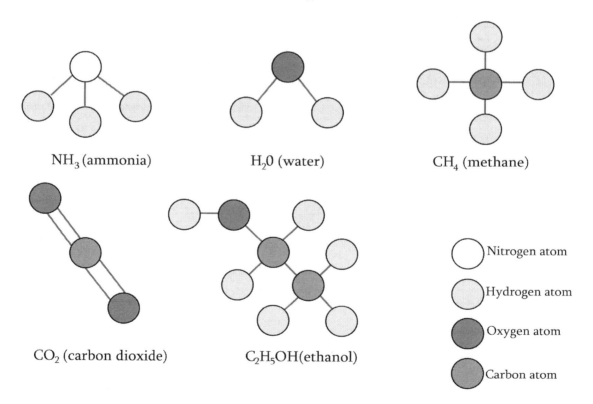

NH_3 (ammonia) H_2O (water) CH_4 (methane)

CO_2 (carbon dioxide) C_2H_5OH(ethanol)

Nitrogen atom
Hydrogen atom
Oxygen atom
Carbon atom

☞ **Take it Further** – For a challenge, see how long of a polyethylene chain you can make, using the marshmallows and rules set forth in the experiment (***Hint***—*Polyethylene is composed of only carbon and hydrogen.*)

Discussion Questions

1. What is synthesis? (*UIDS pg. 122 - Synthesis is when chemists build larger and more useful compounds from elements or smaller compounds.*)

2. How are compounds and mixtures alike? (*UIDS pg. 122 - Compounds and mixtures are both a combination of different elements.*) How are they different? (*UIDS pg. 122 - Compounds have their own unique chemical properties different from the elements that make them up, while compounds or elements in a mixture still retain their original chemical properties. A mixture can be easily separated, while compounds must be chemically separated.*)

3. What does homogenous mean? Give an example of a homogenous substance. (*UIDS pg. 123 - Homogenous means that a substance has all the particles are in the same phase. In other words, the chemical and physical properties are the same throughout the substance. A solution is an example of a homogenous substance.*)

4. What does heterogenous mean? Give an example of a heterogenous substance. (*UIDS pg. 123 - Heterogenous means that a substance has particles that are in different phases. In other words, the chemical and physical properties can be different throughout the substance. A suspension is an example of a heterogenous substance.*)

5. How do chemists define the purity of a substance? (*UIDS pg. 123 - Chemists say that a substance is pure when it contains only one type of element or compound.*)

Want More

↻ **Miscible or Immiscible** – Have the students gather water, oil, juice, alcohol and food coloring from your kitchen pantry or fridge. Then, have them mix each of the liquids with one of the other liquids to determine if they are miscible or immiscible. (*This concept will be touched on again next week.*)

Student Assignment Sheet Week 11
Solutions

Experiment: Is it polar or nonpolar?
Materials:

- ✓ 5 clear cups (or beakers)
- ✓ 5 plastic spoons
- ✓ Sugar
- ✓ Salt
- ✓ Baking soda
- ✓ Flour
- ✓ Petroleum jelly
- ✓ Water
- ✓ Vegetable oil
- ✓ Tablespoon

Procedure:

1. Read the introduction to this experiment and write down which molecules you think are polar and which ones you think are nonpolar.
2. Label the cups #1 through #5. Add ½ cup (120 mL) of room temperature water to each of the five cups.
3. Then stir in 1 TBSP (12.6 g) of sugar to cup #1, 1 TBSP (17.1 g) of salt to cup #2, 1 TBSP (12 g) of baking soda to cup #3, 1 TBSP (5 g) of flour to cup #4 and 1 TBSP (8 g) of petroleum jelly to cup #5.
4. Gently stir each of the cups with a spoon, using a different spoon for each cup. Wait 10 minutes and record your observations and results.
5. Pour out the solutions, rinse well and dry each of the cups. Add ½ cup (120 mL) of room temperature oil to each of the five cups.
6. Repeat steps 3 and 4.
7. Draw conclusions and complete your experiment sheet.

Vocabulary & Memory Work

- ☐ Vocabulary: solution, solubility, solute, solvent
- ☐ Memory Work—This week, add the following elements to what you are working on memorizing:
    - ✓ 41-Nb-Niobium, 42-Mo-Molybdenum, 43-Tc-Technetium, 44-Ru-Ruthenium
- ☐ Memory Work—Continue to work on the Law of Constant Composition

Sketch: Anatomy of a Water Molecule

- 🕐 Label the following: chemical bonds, oxygen atom, hydrogen atoms, the charges associated with the atoms in the molecule

Writing

- ✍ Reading Assignment: *Usborne Illustrated Dictionary of Science* pp. 144-145 (Solutions and Solubility)
- ✍ Additional Research Readings:
    - 📖 Solutions: *KSE* pp. 158-159
    - 📖 Mixtures: *USE* pp. 58-59

Dates

- 🕐 No dates to be entered this week.

Schedules for Week 11
Two Days a Week

Day 1	Day 2
☐ Do the "Is it polar or nonpolar?" experiment, then fill out the experiment sheet on SG pp. 84-85 ☐ Define solution, solubility, solute, and solvent on SG pg. 74 ☐ Enter the dates onto the date sheets on SG pp. 8-13	☐ Read pp. 144-145 from *UIDS,* then discuss what was read ☐ Color and label the "Anatomy of a Water Molecule" sketch on SG pg. 83 ☐ Prepare an outline or narrative summary, write it on SG pp. 86-87

Supplies I Need for the Week
- ✓ 5 clear cups (or beakers), 5 plastic spoons
- ✓ Sugar, salt, baking soda, flour, petroleum jelly
- ✓ Water, vegetable oil
- ✓ Tablespoon

Things I Need to Prepare

Five Days a Week

Day 1	Day 2	Day 3	Day 4	Day 5
☐ Do the "Is it polar or nonpolar?" experiment, then fill out the experiment sheet on SG pp. 84-85 ☐ Enter the dates onto the date sheets on SG pp. 8-13	☐ Read pp. 144-145 from *UIDS,* then discuss what was read ☐ Write an outline on SG pg. 86	☐ Define solution, solubility, solute, and solvent on SG pg. 74 ☐ Color and label the "Anatomy of a Water Molecule" sketch on SG pg. 83	☐ Read one or all of the additional reading assignments ☐ Write a report from what you learned on SG pg. 87	☐ Complete one of the Want More Activities listed **OR** ☐ Study a scientist from the field of Chemistry

Supplies I Need for the Week
- ✓ 5 clear cups (or beakers), 5 plastic spoons
- ✓ Sugar, salt, baking soda, flour, petroleum jelly
- ✓ Water, vegetable oil
- ✓ Tablespoon

Things I Need to Prepare

Additional Information Week 11

Experiment Information

☞ **Introduction** – (*from the Student Guide*) A solution contains several different types of molecules that have been evenly mixed. Generally, a solution is made when one type of molecule, which we call the solute, is dissolved in another, which we call the solvent. The molecules can be either polar or nonpolar. Polar molecules have a positive and a negative end, while nonpolar molecules do not carry a charge. As a general rule, like dissolves like; this means that solvents with polar molecules dissolve solutes with polar molecules and vice versa. In this experiment, you are going to use a polar solvent, water, and a nonpolar solvent, vegetable oil, to determine if several substances are polar or nonpolar.

☞ **Results** – The students should have the following for the results chart:

Solute	Solvent: Water	Solvent: Vegetable Oil
Sugar	SS	NS
Salt	S	NS
Baking Soda	S	NS
Flour	NS	SS
Petroleum Jelly	NS	S

*S = Soluble, SS = Semi-soluble, NS = Not soluble

They should find that salt, baking soda and sugar are polar molecules, and that flour and petroleum jelly are both nonpolar molecules.

☞ **Explanation** – Salt and baking soda have ionic bonds, meaning the atoms have exchanged electrons to form a chemical bond that holds the molecule together through the associated charges. These two molecules are very polar substances which can easily dissolve in a polar solvent. Sugar has a long carbon chain that has no associated charge and an OH group that carry a charge. However, it is polar because the pull of the OH group is stronger than the carbon chain, so sugar dissolves better in water than in oil. Flour is composed of thousands of different molecules that come from grinding up grain. Some of these molecules are polar, some are not. On the whole, cooking flour contains more nonpolar molecules, so it distributes more evenly in the oil. Petroleum jelly is a mixture of hydrocarbons, all of which are nonpolar, which is why it dissolves easily in the oil and not in the water.

☞ **Take if Further** – Test to see if iodine crystals are polar or nonpolar. (***Note***—*If you cannot find iodine crystals to purchase, you can make your own by mixing iodine tincture from the pharmacy with hydrogen peroxide. The crystals will fall to the bottom of the glass and then you can just scoop them out.*) Use the crystals to repeat the experiment. (*The students should see that the iodine crystals are soluble in oil because they are nonpolar molecules.*)

Discussion Questions

1. Why is water such a good solvent for ionic solids? (*UIDS pg. 144 - Water is a good solvent for ionic compounds because water molecules are polar, meaning they have a slight electrical charge that attracts the ions from the ionic solid.*)

2. How do non-polar solvents work? (*UIDS pg. 144 - Non-polar solvents dissolve covalent compounds by pulling the solute molecules away from the lattice, diffusing them between solvent molecules, and holding the solute molecules there using weak Vander Waals forces.*)

3. What is the difference between concentrated and dilute solutions? (*UIDS pg. 144 - Concentrated solutions have a large amount of solute in relation to solvent. Dilute solutions have a small amount of solute in relation to solvent.*)

4. What is a saturated solution? (*UIDS pg. 145 - A solution is said to be saturated when it cannot hold any more of the solute at a given temperature.*)

5. What happens to the solubility of a solid as the temperature increases? (*UIDS pg. 145 - The solubility of a solid increases as the temperature increases.*) Is it the same for a gas? (*UIDS pg. 145 - No, the solubility of a gas decreases as the temperature increases.*)

6. What is a precipitate? (*UIDS pg. 145 - A precipitate is an insoluble solid that forms in a reaction.*)

7. What is a colloid? (*UIDS pg. 145 - A colloid is a mixture of fine particles of different substances that are dispersed evenly and do not dissolve.*)

Want More

☞ **Research Report** – Have the students research more about polar and nonpolar molecules. Then, have them write a one to three paragraph report sharing what they have found.

☞ **Hard Water Test** – Have the students add a cup (240 mL) of tap water to a glass. Then, add a teaspoon (5 mL) of dish liquid and mix well. Next, use a straw to blow into the water. If lots of bubbles form, then the water is not very hard, but if bubbles do not form, you have hard water.

Sketch Week 11

Anatomy of a Water Molecule

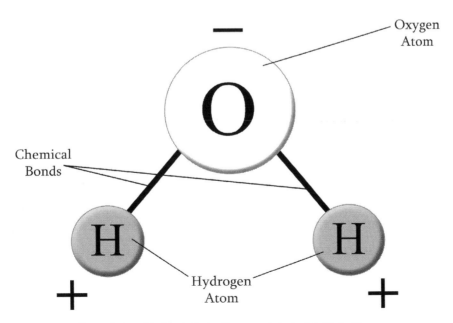

Oxygen Atom

Chemical Bonds

Hydrogen Atom

Chemistry Unit 3: Solutions ~ Week 11 Solutions

Student Assignment Sheet Week 12
Separating Mixtures

Experiment: What colors are in a marker?

Materials:
- ✓ 4 coffee filters
- ✓ Markers
- ✓ Alcohol
- ✓ Coffee can or a wide-mouthed jar
- ✓ Rubber bands
- ✓ Eye dropper

Procedure:
1. Read the introduction to this experiment and make your hypothesis.
2. Choose four different colors of marker to test. In the center of one of the coffee filters, color the outline of a circle with a ½ inch (1.2 cm) diameter with one of the markers you chose. Repeat for the remaining markers.
3. Place one of the colored coffee filters over the top of the coffee can with the circle outline in the center of the opening. Use a rubber band to attach the coffee filter to the can.
4. Then, use the eye dropper to drop 7 to 10 drops of rubbing alcohol in the center of the circle outline on the coffee filter. Wait, observe what happens, and record the colors that are left on the filter paper when the alcohol reaches the edge of the can. (**Note**—*If the alcohol does not reach the edge of the can, you may need to add a drop or two more.*)
5. Repeat steps 3 and 4 for the remaining colored coffee filters.
6. Draw conclusions and complete your experiment sheet.

Vocabulary & Memory Work
- ☐ Vocabulary: chromatography anode, cathode, electrolyte
- ☐ Memory Work—This week, add the following elements to what you are working on memorizing:
  - ✓ 45-Rh-Rhodium, 46-Pd-Palladium, 47-Ag-Silver, 48-Cd-Cadmium
- ☐ Memory Work: Continue to work on the Law of Constant Composition

Sketch: Filtration
- 🖼 Label the Following: filtrate, filter paper that traps residue, suspension of solid in liquid

Writing
- ᐤ Reading Assignment: *Usborne Science Encyclopedia* pp. 60-61 (Separating Mixtures)
- ᐤ Additional Research Readings:
  - 📖 Investigating Substances: *UIDS* pg. 220-221
  - 📖 Separation and Purification: *KSE* pp. 160-161

Dates
- ⏱ 1778-1829 – English chemist Humphrey Davy lives. He is one of the first scientists to see electrolysis in action.
- ⏱ 1832 – Michael Faraday writes a mathematical equation that can be used to calculate the quantity of the separated elements in electrolysis.

Schedules for Week 12
Two Days a Week

Day 1	Day 2
☐ Do the "What colors are in a marker?" experiment, then fill out the experiment sheet on SG pp. 90-91 ☐ Define chromatography on SG pg. 75 ☐ Enter the dates onto the date sheets on SG pp. 8-13	☐ Read pp. 60-61 from *USE,* then discuss what was read ☐ Color and label the "Filtration" sketch on SG pg. 89 ☐ Prepare an outline or narrative summary, write it on SG pp. 92-93

Supplies I Need for the Week
- ✓ 4 coffee filters
- ✓ Set of markers, alcohol
- ✓ Coffee can or a wide-mouthed jar
- ✓ Rubber bands
- ✓ Eye dropper

Things I Need to Prepare

Five Days a Week

Day 1	Day 2	Day 3	Day 4	Day 5
☐ Do the "What colors are in a marker?" experiment, then fill out the experiment sheet on SG pp. 90-91 ☐ Enter the dates onto the date sheets on SG pp. 8-13	☐ Read pp. 60-61 from *USE,* then discuss what was read ☐ Write an outline on SG pg. 92	☐ Define chromatography on SG pg. 75 ☐ Color and label the "Filtration" sketch on SG pg. 89	☐ Read one or all of the additional reading assignments ☐ Write a report from what you learned on SG pg. 93	☐ Complete one of the Want More Activities listed **OR** ☐ Study a scientist from the field of Chemistry

Supplies I Need for the Week
- ✓ 4 coffee filters
- ✓ Set of markers, alcohol
- ✓ Coffee can or a wide-mouthed jar
- ✓ Rubber bands
- ✓ Eye dropper

Things I Need to Prepare

Chemistry Unit 3: Solutions ~ Week 12 Separating Mixtures

Additional Information Week 12

Experiment Information

☞ **Introduction** – (*from the Student Guide*) Scientists can use several different methods to separate a mixture into its components. One of those ways is called chromatography, which uses a solvent to move the molecules of a mixture through a medium, such as filter paper. The smaller the molecule in a mixture, the farther a solvent will carry it. So, chromatography is able to separate the molecules in a mixture according to their size. In this experiment, you are going to use chromatography to separate the different colors found in a marker.

☞ **Results** – The students' results will depend upon the markers that they choose, but in general, the chart below shows what colors the various markers usually show:

Marker Color	Colors Seen
red	red
orange	red, yellow
yellow	yellow
green	yellow, blue
blue	blue
purple	red, blue
black	red, yellow, blue

Some students may also say that they saw orange between the red and yellow, green between the yellow and blue or purple between the red and blue.

☞ **Explanation** – Markers can be made up of several different types of ink; some of those ink molecules are heavier than the others. The alcohol picks up the molecules of ink and carries them along the filter paper. It separates the ink into different colors by depositing the heavier molecules sooner than the lighter ones.

☞ **Take it Further** – Make a colorful chromo-shirt using the same procedure you used in the experiment. The students can use several different colors to make the circle outline to give a more colorful result. They will need to repeat it multiple times to create a full chromo-shirt. Soak the shirt in a salt water solution to set the dye before washing it. (***Note**—Do not wash your chromo-shirt with any other laundry until you are sure that the dye has set.*)

Discussion Questions

1. How do decanting and filtration differ? (*USE pg. 60 - Decanting allows scientists to separate insoluble solids from a liquid by letting the mixture sit. Filtration separates a liquid from a solid by passing the mixture through a filter that allows the liquid to pass, but not the solids.*)
2. Why do scientists use chromatography? (*USE pg. 60 - Scientists use chromatography to analyze the substances in a mixture.*)
3. What is evaporation? (*USE pg. 61 - Evaporation is the process of removing a soluble solid from the solvent by heating the solution.*)
4. When is distillation generally used to separate a solution? (*USE pg. 61 - Distillation is generally used to separate a mixture when the solvent, the liquid part, of the solution is needed.*)

5. How does centrifuging separate a mixture? (*USE pg. 61 - Centrifuging separates liquids and solids in a mixture by spinning them around at a high rate of speed.*)

Want More

⟳ **Filtration** – Have the students make a solution of water and dirt from outside. Then have them line three funnels with several different filtering materials, such as cotton balls, a coffee filter and gravel. Place each of the funnels into a glass cup and have them pour a cup of the dirt solution they made. Observe each of the cups to see which filtration material worked the best.

⟳ **Evaporation** – Have the students make a solution of ½ cup (120 mL) hot water and 1½ TBSP (25.7 g) of salt. Pour the solution onto a plate and set it in on a window sill in the sun. Check on the plate every 30 minutes. The students should see the water evaporate, leaving salt crystals on the plate.

Sketch Week 12

Filtration

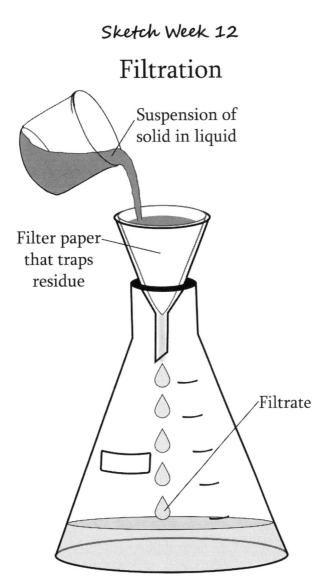

Suspension of solid in liquid

Filter paper that traps residue

Filtrate

Student Assignment Sheet Week 13
Electrolysis

Experiment: Can I split water?

Materials:

- ✓ Distilled water
- ✓ 2 Test tubes
- ✓ Salt
- ✓ Glass cup
- ✓ 2 Alligator clips
- ✓ Covered copper wire
- ✓ 6v Lantern battery
- ✓ Permanent marker

> ☹ **CAUTION**
> **DO NOT DO THIS EXPERIMENT WITHOUT ADULT SUPERVISION!**
> This experiment has the potential to form hydrogen gas, which is extremely flammable.

Procedure:

1. Read the introduction to this experiment and make your hypothesis.
2. Mix together 4 cups (1000 mL) of distilled water with ½ cup (136.5 g) of salt and set the solution aside. Next, cut your wire into two lengths and expose the two ends of each of the wires. Attach one end of the first length of wire to one of the alligator clips. Repeat for the second wire.
3. Now, tape your test tubes to opposite sides of your cup. Fill the cup half of the way full with your salt water solution. Some of the water will go into your test tubes, which is what you want. They should be fairly close to full of water. Fit the other exposed end of your wire into the bottom of your test tubes, making sure that it does not touch the sides of your test tube. Rest the remaining part of the wire on the side or your container
4. Then, mark the water level in each of the tubes with a permanent marker and check your set up to make any necessary adjustments. Now, you can attach the alligator clips to each of the battery terminals of the lantern battery. ***CAUTION—Do NOT touch the water when the wire is attached to the battery!!!***
5. Wait for 30 minutes and then detach the wires from the battery. Mark the water level on each of the tubes once again. Remove the tubes and measure the distance between each of your lines to determine how much gas you collected.
6. Draw conclusions and complete your experiment sheet.

Vocabulary & Memory Work

- ☐ Vocabulary: anode, cathode, electrolyte
- ☐ Memory Work—This week, add the following elements to what you are working on memorizing:
    - ✓ 49-In-Indium, 50-Sn-Tin, 51-Sb-Antimony, 52-Te-Tellurium

Sketch: Electrolytic Cell

- ▣ Label the following: Positive terminal of battery, Electrons leave cell here, Negative terminal of battery, Electrons enter the cell here, Anode is the electrode with a positive charge, Cathode is the electrode with a negative charge, Electrolyte solution

Writing

- ✍ Reading Assignment: *Usborne Illustrated Dictionary of Science* pp. 156-157 (Electrolysis)
- ✍ Additional Research Readings:
    - 📖 Electrolysis: *USE* pp. 82-83

Dates

- ⏱ 1919 – English chemist Francis Ashton invents the mass spectrometer.

Schedules for Week 13
Two Days a Week

Day 1	Day 2
☐ Do the "Can I split water?" experiment, then fill out the experiment sheet on SG pp. 96-97 ☐ Define anode, cathode, and electrolyte on SG pg. 75 ☐ Enter the dates onto the date sheets on SG pp. 8-13	☐ Read pp. 156-157 from *UIDS,* then discuss what was read ☐ Color and label the "Electrolytic Cell" sketch on SG pg. 95 ☐ Prepare an outline or narrative summary, write it on SG pp. 98-99

Supplies I Need for the Week
✓ Distilled water, salt
✓ 2 Test tubes, glass cup or beaker
✓ 2 Alligator clips, covered copper wire, 6-volt Lantern battery
✓ Permanent marker

Things I Need to Prepare

Five Days a Week

Day 1	Day 2	Day 3	Day 4	Day 5
☐ Do the "Can I split water?" experiment, then fill out the experiment sheet on SG pp. 96-97 ☐ Enter the dates onto the date sheets on SG pp. 8-13	☐ Read pp. 156-157 from *UIDS,* then discuss what was read ☐ Write an outline on SG pg. 98	☐ Define anode, cathode, and electrolyte on SG pg. 75 ☐ Color and label the "Electrolytic Cell" sketch on SG pg. 95	☐ Read one or all of the additional reading assignments ☐ Write a report from what you learned on SG pg. 99	☐ Complete one of the Want More Activities listed **OR** ☐ Take the Unit 3 Test

Supplies I Need for the Week
✓ Distilled water, salt
✓ 2 Test tubes, glass cup or beaker
✓ 2 Alligator clips, covered copper wire, 6-volt Lantern battery
✓ Permanent marker

Things I Need to Prepare

Additional Information Week 13

Experiment Information

☞ **Caution – Do NOT allow your students to do this experiment on their own as it has two main hazards, electricity with water and potentially flammable hydrogen gas.** It can be done safely in your home as long as you follow the two rules below:

   1. **Do NOT touch the water when the wire is connected to the battery.** You can wear rubber gloves when you are dealing with the electrical part of this experiment for an extra measure of safety.

   2. **Do NOT do this experiment anywhere near an open flame.** Also, do not smoke in the house while performing this experiment.

☞ **Introduction** – (*from the Student Guide*) Scientists can use a variety of methods to separate mixtures, such as filtration, decantation, distillation, evaporation and centrifugation. Electrolysis separates the components of a molecule within a solution. It uses electrical current to break the compound into its pieces. This process requires that the substance can conduct electricity and that it is either in solution or in a molten state. Electricity passes from two electrodes into the solution, which breaks the compound into its parts. In this experiment we are going to see if we can use electrolysis to split water.

☞ **Results** – The students should see gas collected in each test tube. They should see twice as much gas in the cathode test tube, which is connected to the positive terminal on their battery.

☞ **Explanation** – We added salt to the water so that it would conduct electricity and allow the process of electrolysis to occur. The bubbles you saw at the tip of the wires in the solution were beads of oxygen and hydrogen gas. These two elements make up the compound water. The oxygen gas forms at the anode, which is the wire connected to the negative terminal on your battery. The hydrogen gas at the cathode, which is the wire connected to the positive terminal on your battery. You were able to collect twice as much hydrogen gas than oxygen gas because there are twice as many hydrogen atoms in water than there are oxygen atoms.

☞ **Take it Further** – Test to see if the concentration of salt in the water produces more or less of the two gases. Make sure that you use distilled water to make up your salt water solutions. (*The students should see that the more salt they added to the water, the quicker the electrolysis occurred.*)

Discussion Questions

   1. What is an electrolyte, and why do these substances conduct electricity? (*UIDS pg. 156 - An electrolyte is a compound that conducts electricity in a solution or when it is molten. This is because the ions from the compound are free to move. Cations in an electrolyte carry a positive charge, while anions in an electrolyte carry a negative charge.*)

   2. What is an electrode? (*UIDS pg. 156 - An electrode is a piece of metal or graphite that allows the electrical current to flow into the electrolyte.*)

   3. What is electrorefining? (*UIDS pg. 157 - Electrorefining is the process of using electrolysis to purify a metal such as copper.*)

   4. How does electroplating use electrolysis? (*UIDS pg. 157 - Electroplating uses electrolysis to coat a given object with a thin layer of a metal.*)

Want More

☞ **Research Report** – Have the students research more about how electrolysis is used in industry. Then, have them write a one to three paragraph report sharing what they have found.

Sketch Week 13

Electrolytic Cell

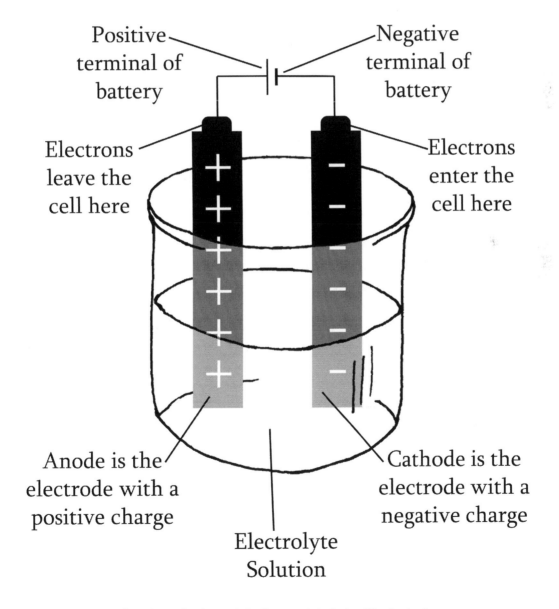

Positive terminal of battery

Negative terminal of battery

Electrons leave the cell here

Electrons enter the cell here

Anode is the electrode with a positive charge

Cathode is the electrode with a negative charge

Electrolyte Solution

Unit 3: Solutions
Unit Test Answers

Vocabulary Matching

1. D
2. C
3. A
4. G

5. B
6. H
7. K
8. I

9. J
10. L
11. F
12. E

True or False

1. False (*A mixture is composed of elements or compounds that have not been chemically joined.*)
2. True
3. True
4. True

5. False (*Evaporation is a way to separate a soluble solid that has been dissolved in a liquid solvent.*)
6. True
7. True
8. True

Short Answer

1. The law of constant composition states that a pure compound always contains the same elements in the same proportions.
2. Concentrated solutions have a large amount of solute in relation to solvent. Dilute solutions have a small amount of solute in relation to solvent.
3. Students answers should include two of the following: Evaporation is a way to separate a soluble solid that has been dissolved in a liquid solvent. Filtration is used to separate a suspension of particles of a solid that are spread through a liquid. Centrifuging is used to separate a suspension when filtering or settling will not work. Distillation is used to separate a mixture of two or more liquids. Fractional distillation is used to separate two or more liquids when the liquids have similar boiling points.
4. An electrolyte is a compound that conducts electricity in a solution or when it is molten. This is because the ions from the compound are free to move. Cations in an electrolyte carry a positive charge, while anions in an electrolyte carry a negative charge.
5. 37-Rb-Rubidium, 38-Sr-Strontium, 39-Y-Yttrium, 40-Zr-Zirconium, 41-Nb-Niobium, 42-Mo-Molybdenum, 43-Tc-Technetium, 44-Ru-Ruthenium, 45-Rh-Rhodium, 46-Pd-Palladium, 47-Ag-Silver, 48-Cd-Cadmium, 49-In-Indium, 50-Sn-Tin, 51-Sb-Antimony, 52-Te-Tellurium

Unit 3: Solutions
Unit Test

Vocabulary Matching

1. Molecule ___

2. Mixture ___

3. Miscible ___

4. Immiscible ___

5. Solution ___

6. Solubility ___

7. Solvent ___

8. Solute ___

9. Chromatography ___

10. Anode ___

11. Cathode ___

12. Electrolyte ___

A. Liquids that can be blended together.

B. A homogenous mixture of two or more substances.

C. A combination of two or more elements or compounds that are not chemically combined.

D. A substance that is formed when two or more atoms chemically join together.

E. A substance that conducts electricity when it is in solution.

F. The positively charged electrode by which current enters the cell.

G. Liquids that cannot be blended together.

H. The ability of a solute to be dissolved.

I. The substance that dissolves in the solvent to form a solution.

J. A method of separating the substances in a mixture by the rate they move through or along a medium, such as filter paper.

K. The substance in which the solute dissolves to form a solution, typically a liquid.

L. The negatively charged electrode by which current leaves the cell.

True or False

1. _____ A mixture is composed of elements or compounds that have been chemically joined.

2. _____ Chemists say that a substance is pure when it contains only one element or compound.

3. _____ Water is a good solvent because it is a polar molecule.

96

4. _____ The solubility of a solid increases as the temperature increases. The solubility of a gas decreases as the temperature increases.

5. _____ Evaporation is the easiest method for separating a suspension.

6. _____ Distillation is a method of separating a mixture that it best used when you need to separate the solvent from the solution.

7. _____ An electrode is a piece of metal or graphite that allows the electrical current to flow into the electrolyte.

8. _____ Electrorefining is the process of using electrolysis to purify a metal, such as copper.

Short Answer

1. What is the law of constant composition?

2. What is the difference between concentrated and dilute solutions?

3. Name two types of separating a mixture and why you would choose each of the methods.

4. What is an electrolyte, and why do these substances conduct electricity?

5. Fill in elements 37-52 from the periodic table.

- Rubidium
- Tin
- Yttrium
- Ruthenium
- Zirconium
- Niobium
- Strontium
- Molybdenum
- Cadmium
- Rhodium
- Palladium
- Tellurium
- Silver
- Indium
- Technetium
- Antimony

37. _____

38. _____

39. _____

Chemistry Unit 3: Solutions ~ Unit Test

40. _____

41. _____

42. _____

43. _____

44. _____

45. _____

46. _____

47. _____

48. _____

49. _____

50. _____

51. _____

52. _____

Chemistry: Unit 4

Chemical Reactions

Unit 4: Chemical Reactions
Overview of Study

Sequence of Study

Week 14: Chemical Bonding
Week 15: Chemical Reactions – Part 1
Week 16: Chemical Reactions – Part 2
Week 17: Chemical Reactions – Part 3
Week 18: Catalysts
Week 19: Oxidation and Reduction

Materials by Week

Week	Materials
14	Cake frosting, Red and yellow bite-sized candies (such as regular sized M&M's)
15	Yeast, Hydrogen peroxide, Epsom salts, Water, 2 cups, 2 thermometers
16	Baking soda, Chalk, Iron nail (non-coated), Copper penny, White vinegar, 4 cups
17	2 potatoes, Pot, Water, Oven mitt, Large Slotted Spoon
18	Carbonated water, Sugar, 2 cups
19	Steel wool, Vinegar, Jar with lid, Ammonia, Hydrogen peroxide

Vocabulary for the Unit

1. **Ion** – An atom or group of atoms that has become charged by gaining or losing one or more electrons.
2. **Ionic Bonding** – A strong chemical bond that is formed by the attraction between two ions of opposite charges.
3. **Covalent Bonding** – A chemical bond between two atoms, in which each atom shares an electron.
4. **Metallic Bonding** – A chemical bond where positive metal ions form a lattice structure with freely moving electrons between them.
5. **Chemical Reaction** – An interaction between the atoms of two substances where the atoms rearrange to form two new molecules.
6. **Endothermic Reaction** – A chemical reaction that takes in heat.
7. **Exothermic Reaction** – A chemical reaction that gives off heat.
8. **Product** – A new substance produced in a chemical reaction.
9. **Reactant** – The substance present at the beginning of a chemical reaction.
10. **Moles** – The SI unit for the amount of a substance.
11. **Reactivity** – The tendency of a substance to react with other substances.

12. **Concentration** – A measure of the strength of a solution, i.e., the amount of the solute in the solvent.
13. **Equilibrium** – The point at which the rate of the forward reaction is equal to the rate of the backward reaction.
14. **Irreversible Reaction** – A reaction in which the products can never be turned back into the reactants.
15. **Reversible Reaction** – A reaction in which, under the right conditions, the products can be turned back into the reactants.
16. **Activation Energy** – The minimum amount of energy a reaction needs to start.
17. **Catalyst** – A substance which speeds up a chemical reaction without being changed by the reaction.
18. **Oxidation** – Occurs when an atom loses electrons in a reaction, or when a substance loses a hydrogen or gains an oxygen atom in a chemical reaction.
19. **Reduction** – Occurs when an atom gains electrons in a reaction, or when a substance gains a hydrogen or loses an oxygen atom in a chemical reaction.
20. **Redox Reaction** – A reaction that primarily involves the transfer of electrons between two substances.

Memory Work for the Unit

The Elements of the Periodic Table – The following elements will be memorized in this unit:

- ✓ 53-I-Iodine
- ✓ 54-Xe-Xenon
- ✓ 55-Cs-Cesium
- ✓ 56-Ba-Barium
- ✓ 57-La-Lanthanum
- ✓ 58-Ce-Cerium
- ✓ 59-Pr-Praseodymium
- ✓ 60-Nd-Neodymium
- ✓ 61-Pm-Promethium
- ✓ 62-Sm-Samarium
- ✓ 63-Eu-Europium
- ✓ 64-Gd-Gadolinium
- ✓ 65-Tb-Terbium
- ✓ 66-Dy-Dysprosium
- ✓ 67-Ho-Holmium
- ✓ 68-Er-Erbium
- ✓ 69-Tm-Thulium
- ✓ 70-Yb-Ytterbium
- ✓ 71-Lu-Lutetium
- ✓ 72-Hf-Hafnium
- ✓ 73-Ta-Tantalum
- ✓ 74-W-Tungsten
- ✓ 75-Re-Rhenium
- ✓ 76-Os-Osmium

Law of Conservation of Mass – In a chemical reaction, the mass of the products is equal to the mass of the reactants.

Avogadro's number: 6.02×10^{23} (This tells us the number of particles per mole.)

Le Chatelier's Principle – If a change is made to a reaction in equilibrium, the reaction will adjust itself to counter the effects of that change.

Student Assignment Sheet Week 14
Chemical Bonding

Experiment: Chocolate Bonding

Materials:
- ✓ Cake frosting
- ✓ Red and yellow bite-sized candies (such as regular sized M&M's)

Procedure:
1. Read the introduction to this experiment.
2. **Ionic bonding** – Count out nine red candies and one yellow candy. Arrange the nine red candies in a circle, so that there are two in the center with seven surrounding them. (*Be sure to leave a space for one more in the outer circle.*) These are the electrons in the outer shell of a chlorine atom. Take the one yellow candy, which represents the electron in the outer shell of a sodium atom, and move into the space in the outer circle of the chlorine electrons. Now you have an ionic bond between your sodium and chlorine atom. Draw a picture of your model or take a picture of it for your experiment sheet.
3. **Covalent bonding** – Count out eight red candies and two yellow candies. Use the eight red candies to make a model showing the electrons in the outer shell of an oxygen atom by making a circle with six candies on the outside and two on the inside. (*Be sure to leave space for two more candies in the outer layer.*) Now take two of the yellow candies, which represent the electrons in the outer shells of the two different hydrogen atoms, and bite or cut them in half. Do the same with the two of the red candies from the outer circle you created. Then, use the frosting to glue together one half of each color to form four whole candies. Place the dual colored candies in the outer circle of your model to represent the covalent bond that is found between oxygen and hydrogen in water. Draw a picture of your model or take a picture of it and then complete your experiment sheet.

Vocabulary & Memory Work
- ☐ Vocabulary: covalent bonding, ion, ionic bonding, metallic bonding
- ☐ Memory Work—This week, add the following elements to what you are working on memorizing:
  - ✓ 53-I-Iodine, 54-Xe-Xenon, 55-Cs-Cesium, 56-Ba-Barium

Sketch: Types of Chemical Bonding
- 🖼 Label the following on the Ionic Bond: donated electron, sodium ion, chlorine ion
- 🖼 Label the following on the Covalent Bond: shared electrons, oxygen atom, hydrogen atom
- 🖼 Label the following on the Metallic Bond: fixed positive gold ions, free electrons

Writing
- ꙮ Reading Assignment: *Usborne Illustrated Dictionary of Science* pg. 130 (Bonding), pg. 131 (Ionic Bonding), pg. 132 (Covalent Bonding), pg. 134 (Metallic Bonding)
- ꙮ Additional Research Readings:
  - 📖 Bonding and Valency: *KSE* pp. 166-167, Bonding: *USE* pp. 68-71

Dates
- ⏲ 1704 – Isaac Newton roughly outlines atomic theory.
- ⏲ 1916 – Gilbert Lewis develops the concept of electron pair bonding.
- ⏲ 1954 – Linus Pauling is awarded the Nobel Prize for Chemistry for his work on calculating the energy to break bonds.

Chemistry Unit 4: Chemical Reactions ~ Week 14 Chemical Bonding

Schedules for Week 14
Two Days a Week

Day 1	Day 2
☐ Do the "Chocolate Bonding" experiment, then fill out the experiment sheet on SG pp. 108-109 ☐ Define ion, ionic bonding, covalent bonding, and metallic bonding on SG pg. 102 ☐ Enter the dates onto the date sheets on SG pp. 8-13	☐ Read pp. 130-132,134 from *UIDS,* then discuss what was read ☐ Color and label the "Types of Chemical Bonding" sketch on SG pg. 107 ☐ Prepare an outline or narrative summary, write it on SG pp. 110-111

Supplies I Need for the Week
- ✓ Cake frosting
- ✓ Red and yellow bite-sized candies (such as regular sized M&M's)

Things I Need to Prepare

Five Days a Week

Day 1	Day 2	Day 3	Day 4	Day 5
☐ Do the "Chocolate Bonding" experiment, then fill out the experiment sheet on SG pp. 108-109 ☐ Enter the dates onto the date sheets on SG pp. 8-13	☐ Read pp. 130-132,134 from *UIDS,* then discuss what was read ☐ Write an outline on SG pg. 110	☐ Define ion, ionic bonding, covalent bonding, and metallic bonding on SG pg. 102 ☐ Color and label the "Types of Chemical Bonding" sketch on SG pg. 107	☐ Read one or all of the additional reading assignments ☐ Write a report from what you learned on SG pg. 111	☐ Complete one of the Want More Activities listed **OR** ☐ Study a scientist from the field of Chemistry

Supplies I Need for the Week
- ✓ Cake frosting
- ✓ Red and yellow bite-sized candies (such as regular sized M&M's)

Things I Need to Prepare

Additional Information Week 14

Experiment Information

☞ **Introduction** – (*from the Student Guide*) There are three main ways that atoms can chemically bond to form molecules. The first is called ionic bonding. In this type of bonding, one atom donates an electron to another to form the bond. The second way is called covalent bonding. In this type of bonding, one atom shares an electron with another atom to form the bond. In this experiment, you are going to use candy to demonstrate both these types of chemical bonding. The third type of bonding is metallic bonding, which we will discuss later.

☞ **Results** – This experiment is meant to be a demonstration of how chemical bonding works so that the concepts will be solidified in your student's mind. The picture below will give you an idea of what their creations should look like:

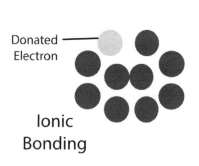
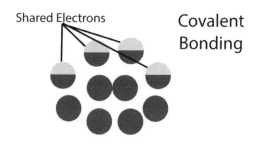

Discussion Questions

1. What does all bonding involve? (*UIDS pg. 130 - All bonding involves the losing or sharing of electrons from the outermost shells of an atom.*)
2. What is a valency electron? (*UIDS pg. 130 - A valency electron is an electron in the outer shell of an atom that is used in the forming of a bond.*)
3. What is a cation? (*UIDS pg. 130 - A cation is an ion that is formed when an atom loses an electron in a reaction, resulting in a positive charge.*) An anion? (*UIDS pg. 130 - An anion is an ion that is formed when an atom gains an electron in a reaction, resulting in a negative charge.*)
4. What happens when an ionic bond forms? (*UIDS pg. 131 - When an ionic bond happens, one atom loses electrons, while the other gains them, which forms oppositely charged ions that are attracted to each other. This attraction causes the atoms to stay together.*)
5. How is a covalent bond formed? (*UIDS pg. 132 - A covalent bond is formed when two atoms share electrons between each other so that each atom has a stable outer shell. One pair of shared electrons makes a single covalent bond.*)
6. What is metallic bonding? (*UIDS pg. 134 - Metallic bonding occurs when the electrons from the outer shell of a metal atom float around in a "sea" of electrons from other metal atoms.*)
7. What is a hydrogen bond? (*A hydrogen bond occurs between polar molecules that have a hydrogen and a lone pair of electrons in another molecule.*)

Chemistry Unit 4: Chemical Reactions ~ Week 14 Chemical Bonding

Want More

⟡ **Nomenclature** – Nomenclature, or naming compounds in chemistry, came about when scientists wanted a consistent way to refer to the compounds they were discovering. Use the worksheet on pg. 259-260 of the Appendix to teach nomenclature to your student. It contains a very basic explanation of this concept; your student will learn quite a bit more about naming compounds in high school. For now, your goal is to give them a basic understanding of how scientists name chemical compounds.

Answers to worksheet practice:

- ✓ **Ionic Compounds:** MgO - magnesium oxide, NaBr - sodium bromide, Li_2S - lithium sulfide, $MgSO_4$ - magnesium sulfate, $Be(OH)_2$ - beryllium hydroxide, $Sr(NO_3)_2$ - strontium nitrate

- ✓ **Covalent Compounds:** CO_2 - carbon dioxide, NO_2 - nitrogen dioxide, SO_3 - sulfur trioxide, N_2S - dinitrogen sulfide, BF_3 - boron trifluoride, P_2Br_4 - diphosphorus tetrabromide

Sketch Week 14

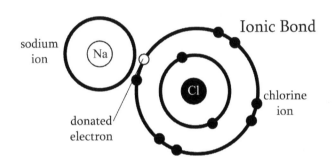

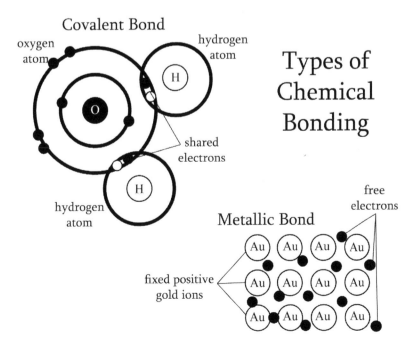

Chemistry Unit 4: Chemical Reactions ~ Week 14 Chemical Bonding

Student Assignment Sheet Week 15
Chemical Reactions – Part 1

Experiment: Exothermic or Endothermic?
  Materials:
  - ✓ Yeast
  - ✓ Hydrogen peroxide
  - ✓ Epsom salts
  - ✓ Water
  - ✓ 2 cups
  - ✓ 2 thermometers

  Procedure:
  1. Read the introduction to this experiment and circle what you think the reactions will be.
  2. Add 1 cup (240 mL) of hydrogen peroxide to one of the cups and 1 cup (240 mL) of warm water to the other. Place a thermometer in each up and wait 5 minutes. Record the temperature of both cups as the starting temperature.
  3. Now, add 1 TBSP (8.5 g) of yeast to the hydrogen peroxide and gently stir. Then, add 1 TBSP (14.2 g) of Epsom salts to the water and gently stir. Wait until the temperature on the thermometer stabilizes and record the final temperature for the reaction.
  4. Draw conclusions and complete your experiment sheet.

Vocabulary & Memory Work
  - ☐ Vocabulary: chemical reaction, endothermic reaction, exothermic reaction, product, reactant
  - ☐ Memory Work—This week, add the following elements to what you are working on memorizing:
    - ✓ 57-La-Lanthanum, 58-Ce-Cerium, 59-Pr-Praseodymium, 60-Nd-Neodymium
  - ☐ Memory Work—Work on memorizing the Law of Conservation of Mass – In a chemical reaction, the mass of the products is equal to the mass of the reactants.

Sketch: Chemical Reaction
  - ▨ Label the following: hydrogen, oxygen, oxygen atoms, hydrogen atoms, water

Writing
  - ✍ Reading Assignment: *Usborne Science Encyclopedia* pp. 76-77 (Chemical Reactions)
  - ✍ Additional Research Readings:
    - 📖 Energy and Chemical Reactions: *UIDS* pp. 162-163
    - 📖 Chemical Reactions: *KSE* pp. 162-163

Dates
  - 🕐 1620 – Francis Bacon publishes a book, *New Method*, in which he states that theories need to be supported by experiments.
  - 🕐 1661 – Robert Boyle publishes *The Sceptical Chymist*, in which he says ideas should always be tested through experiments.
  - 🕐 1789 – Antoine Lavoisier establishes that matter cannot be created or destroyed.

Schedules for Week 15
Two Days a Week

Day 1	Day 2
☐ Do the "Exothermic or Endothermic?" experiment, then fill out the experiment sheet on SG pp. 114-115 ☐ Define chemical reaction, endothermic reaction, exothermic reaction, product, and reactant on SG pg. 102 ☐ Enter the dates onto the date sheets on SG pp. 8-13	☐ Read pp. 76-77 from *USE,* then discuss what was read ☐ Color and label the "Chemical Reaction" sketch on SG pg. 113 ☐ Prepare an outline or narrative summary, write it on SG pp. 116-117

Supplies I Need for the Week
- ✓ Yeast, Hydrogen peroxide
- ✓ Epsom salts, Water
- ✓ 2 cups, 2 thermometers

Things I Need to Prepare

Five Days a Week

Day 1	Day 2	Day 3	Day 4	Day 5
☐ Do the "Exothermic or Endothermic?" experiment, then fill out the experiment sheet on SG pp. 114-115 ☐ Enter the dates onto the date sheets on SG pp. 8-13	☐ Read pp. 76-77 from *USE,* then discuss what was read ☐ Write an outline on SG pg. 116	☐ Define chemical reaction, endothermic reaction, exothermic reaction, product, and reactant on SG pg. 102 ☐ Color and label the "Chemical Reaction" sketch on SG pg. 113	☐ Read one or all of the additional reading assignments ☐ Write a report from what you learned on SG pg. 117	☐ Complete one of the Want More Activities listed **OR** ☐ Study a scientist from the field of Chemistry

Supplies I Need for the Week
- ✓ Yeast, Hydrogen peroxide
- ✓ Epsom salts, Water
- ✓ 2 cups, 2 thermometers

Things I Need to Prepare

Additional Information Week 15

Experiment Information

☞ **Introduction** – (*from the Student Guide*) In a chemical reaction, bonds are broken and reformed. This causes the reactants, or the substances you begin with, to be turned into new substances known as the products. Every chemical reaction involves energy that is either released or absorbed in the form of heat, light, or electricity. We call the heat-releasing reactions exothermic reactions and the heat-absorbing reactions are referred to as endothermic reactions. In this experiment, you are going to look at two different reactions to determine if they are exothermic or endothermic.

☞ **Results** – The yeast and hydrogen peroxide reaction is exothermic and the Epsom salts and water reaction is endothermic.

☞ **Explanation** – Hydrogen peroxide is a very powerful oxidizing agent and it immediately begins to breakdown the cell walls of the yeast. This process releases energy into the solution in the form of heat. On the other hand, it requires energy to break the bonds of the magnesium sulfate in the Epsom salts. The energy is absorbed by the magnesium and sulfate ions and removed from the water, which cause the temperature to drop.

☞ **Take it Further** – Try mixing laundry detergent and water or baking soda and water to see if they are exothermic or endothermic reactions. (*Both of these reactions are exothermic.*)

Discussion Questions

1. What are the substances in a reaction called? (*USE pg. 76 - The substances present at the beginning of the reaction are the reactants. The substances present at the end of the reaction are the products.*)

2. What happens in a chemical reaction, and what does it require? (*USE pg. 76 - In a chemical reaction, bonds are broken and made. This process requires energy.*)

3. What is the difference between exothermic and endothermic reactions? (*USE pg. 76 - In an endothermic reaction, the heat energy needed to break the bonds of the reactants is greater than the heat energy given off when the new bonds of the products are formed. In an exothermic reaction, the heat energy needed to break the bonds of the reactants is less than the heat energy given off when the new bonds of the products are formed.*)

4. What is activation energy? (*USE pg. 76 - Activation energy is the amount of energy a reaction needs to start.*)

5. What is the law of conservation of mass? (*USE pg. 77 - The law of conservation of mass says that matter cannot be created or destroyed during a chemical reaction.*)

6. How do chemists show reactions? (*USE pg. 77 - Chemists show reactions with equations that have the reactants on the right and the products on the left.*)

7. How do chemists measure chemical substances? (*USE pg. 77 - Chemists measure chemical substances using moles.*)

Want More

✐ **Chemical vs. Physical Changes** – Have the students think of some changes that they see every day, such as water being boiled and making a cup of tea. Have them list these changes and then decide if they are chemical or physical changes. (***Note***—*Remember that a chemical change looks different and it is usually a permanent irreversible change. A physical change can look different, but chemically it is still the same. It is not permanent and can be reversed.*) Once they have at least five of each, have them create a poster depicting the difference between chemical and physical changes.

Sketch Week 15

Chemical Reaction

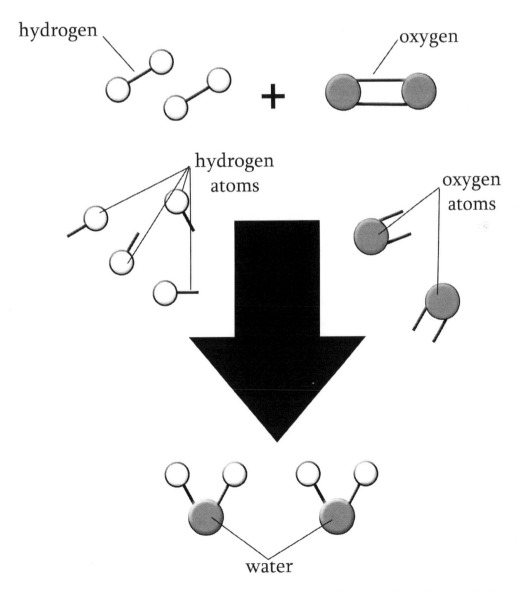

water

Student Assignment Sheet Week 16
Chemical Reactions – Part 2

Experiment: How reactive is it?
> Materials:
> - ✓ Baking soda
> - ✓ Chalk
> - ✓ Iron nail (non-coated)
> - ✓ Copper penny
> - ✓ White vinegar
> - ✓ 4 cups
>
> Procedure:
> 1. Read the introduction to this experiment and write what you think the order the four reactants will be .
> 2. Pour ½ cup (120 mL) of vinegar into each of the 4 cups.
> 3. Add 1 tsp (3.5 g) of baking soda to the first cup. Observe and record how long it takes for a reaction to occur.
> 4. Repeat this process with 1 piece of chalk, 1 iron nail, and 1 copper penny in the other 3 cups. Observe and record how long it takes for a reaction to occur.
> 5. Draw conclusions and complete your experiment sheet.

Vocabulary & Memory Work

- ☐ Vocabulary: moles, reactivity
- ☐ Memory Work—This week, add the following elements to what you are working on memorizing:
  - ✓ 61-Pm-Promethium, 62-Sm-Samarium, 63-Eu-Europium, 64-Gd-Gadolinium
- ☐ Memory Work—Work on memorizing the Avogadro's number: 6.02×10^{23} and what it tells us, along with the Law of Conservation of Mass

Sketch: Anatomy of a Chemical Equation

- ▦ Label the following: reactants, products, numbers are used to balance the equation, symbols show what state the chemical is in, arrow to show the overall flow of the reaction

Writing

- ⤶ Reading Assignment: *Usborne Illustrated Dictionary of Science* pg. 141 (section on Equations) and pp. 158-159 (Reactivity)
- ⤶ Additional Research Readings:
  - 📖 Moles: *UIDS* pg. 139
  - 📖 Reactivity Series: *UIDS* pg. 211

Dates

- 🕐 1811 – Italian scientist Amedeo Avogadro first proposes that the volume of a gas at a constant pressure and temperature is proportional to the number of atoms or molecules regardless of the nature of the gas.
- 🕐 1926 – Jean Perrin wins the Nobel Prize in Physics for determining the constant that Avogadro first proposed. He named the constant Avogadro's number in honor of the scientist.

Chemistry Unit 4: Chemical Reactions ~ Week 16 Chemical Reactions – Part 2

Schedules for Week 16
Two Days a Week

Day 1	Day 2
☐ Do the "How reactive is it?" experiment, then fill out the experiment sheet on SG pp. 120-121 ☐ Define moles and reactivity on SG pg. 103 ☐ Enter the dates onto the date sheets on SG pp. 8-13	☐ Read pp. 141, 158-159 from *UIDS*, then discuss what was read ☐ Label the "Anatomy of a Chemical Equation" sketch on SG pg. 119 ☐ Prepare an outline or narrative summary, write it on SG pp. 122-123

Supplies I Need for the Week
- ✓ Baking soda, Chalk
- ✓ Iron nail (non-coated), Copper penny
- ✓ White vinegar, 4 cups

Things I Need to Prepare

Five Days a Week

Day 1	Day 2	Day 3	Day 4	Day 5
☐ Do the "How reactive is it?" experiment, then fill out the experiment sheet on SG pp. 120-121 ☐ Enter the dates onto the date sheets on SG pp. 8-13	☐ Read pp. 141, 158-159 from *UIDS*, then discuss what was read ☐ Write an outline on SG pg. 122	☐ Define moles and reactivity on SG pg. 103 ☐ Label the "Anatomy of a Chemical Equation" sketch on SG pg. 119	☐ Read one or all of the additional reading assignments ☐ Write a report from what you learned on SG pg. 123	☐ Complete one of the Want More Activities listed **OR** ☐ Study a scientist from the field of Chemistry

Supplies I Need for the Week
- ✓ Baking soda, Chalk
- ✓ Iron nail (non-coated), Copper penny
- ✓ White vinegar, 4 cups

Things I Need to Prepare

Chemistry Unit 4: Chemical Reactions ~ Chemical Reactions – Part 2

Additional Information Week 16

Experiment Information

☞ **Introduction** – (*from the Student Guide*) The reactivity series tells how reactive a given metal can be in its elemental state. Those at the top of the reactivity series, such as potassium, change quickly when exposed to air, while those at the bottom, such as gold, do not change at all. We can also use the periodic table to help us determine an element's propensity to react. For metals, reactivity decreases as you move from left to right on the periodic table and increases and you move down the table. In this experiment, you are going to test the reactivity of sodium (*baking soda*), calcium (*chalk*), iron (*nail*) and copper (*penny*) in vinegar.

☞ **Results** – The students should see that the baking soda reacted immediately and vigorously. The chalk should have begun to bubble lightly fairly soon, but it took hours to dissolve fully. They should see that the iron nail begins to rust after several hours. Finally, the penny should have shown little to no change after several hours in the vinegar.

☞ **Explanation** – Sodium is very high on the reactivity series, which explains why the baking soda, a.k.a. sodium bicarbonate, reacted quickly with the vinegar. Calcium is still relatively high on the series, but not nearly as much as sodium. This is why you saw a reaction beginning right away, but that reaction took much longer to complete than the one with sodium. Iron is even further down on the reactivity series, which explains why the nail took longer to rust. Finally, copper is the lowest metal on the series of the elements we tested, which explains why it took the longest to react.

☞ **Take it Further** – Try the experiment again with aluminum foil to see how the reactivity of aluminum compares to the other four metals you used in the experiment. (*The aluminum foil should eventually turn greyish-black. This reaction should be quicker than the iron reaction, but slower than the calcium reaction.*)

Discussion Questions

1. What does a balanced chemical equation show? (*UIDS pg. 141 - A balanced chemical equation shows the substances involved in the reaction and the proportions in which they need to be mixed.*)
2. What is a spectator ion? (*UIDS pg. 141 - A spectator ion remains the same before and after a chemical reaction.*)
3. What does the reactivity of an element depend on? (*UIDS pg. 158 - The reactivity of an element depends on its ability to lose or gain an electron.*)
4. What is a displacement reaction? (*UIDS pg. 158 - In a displacement reaction the metal ion that is higher on the reactivity series displaces another metal that is lower on the reactivity series.*)
5. What does the reactivity series tell us? (*UIDS pg. 158 - The reactivity series tell us how reactive certain metals are. The metals at the top are very reactive, while those at the bottom are the least reactive.*)
6. What does the electrochemical series tell us? (*UIDS pg. 159 - The electrochemical series tell us whether a metal tends to form a negative or positive ion in an aqueous solution. The more negative the electrode potential, the higher the chance the metal will form a positive ion.*)

Want More

- **Mole Day** – Celebrate Avogadro's number by having your own mole day! See the following website for ideas of activities you could do:
  - 🖥 http://www.moleday.org/
- **Chemical Equations: Part 1** – Use the worksheet on pp. 261-262 of the Appendix to teach the students the basics of balancing chemical equations.

  Answers to part 1 of the worksheet:
  1. $2\ AgNO_3 + Cu \longrightarrow Cu(NO_3)_2 + 2\ Ag$
  2. $AlBr_3 + 3\ K \longrightarrow 3\ KBr + Al$
  3. $2\ LiNO_3 + CaBr_2 \longrightarrow Ca(NO_3)_2 + 2\ LiBr$
  4. $PbBr_2 + 2\ HCl \longrightarrow 2\ HBr + PbCl_2$
  5. $2\ NaCN + CuCO_3 \longrightarrow Na_2CO_3 + Cu(CN)_2$

Sketch Week 16

Anatomy of a Chemical Equation

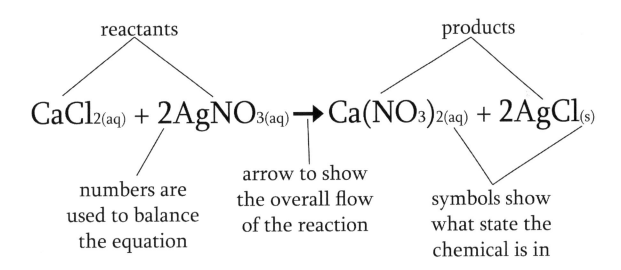

reactants

products

$$CaCl_{2(aq)} + 2AgNO_{3(aq)} \rightarrow Ca(NO_3)_{2(aq)} + 2AgCl_{(s)}$$

numbers are used to balance the equation

arrow to show the overall flow of the reaction

symbols show what state the chemical is in

Student Assignment Sheet Week 17
Chemical Reactions – Part 3

Experiment: Does the surface area affect cooking time?
Materials:
- ✓ 2 potatoes
- ✓ Pot
- ✓ Water
- ✓ Oven mitt
- ✓ Large Slotted Spoon

> ☣ **CAUTION**
>
> **DO NOT DO THIS EXPERIMENT WITHOUT ADULT SUPERVISION!**
> Boiling hot water can cause severe damage. Be sure to use the proper safety gear.

Procedure:
1. Read the introduction to this experiment and answer the question.
2. Take one potato and cut it in half. Then, cut one thin slice off of one of the halves and cut the other half in half. This will give you four sample potatoes, one full size potato, one half, one quarter and one thin slice.
3. Now, fill a pot three quarters of the way with water and place the four sample potatoes in it. Then, set the pot on a burner and turn the burner on medium high.
4. After the water has been boiling for 5 minutes, use the slotted spoon to remove each of the sample potatoes and check for doneness. If the sample is done, remove it from the pot and record how long it took to cook.
5. Repeat step 4 every 5 minutes until all four potato samples are fully cooked.
6. Draw conclusions and complete your experiment sheet.

Vocabulary & Memory Work
- ☐ Vocabulary: concentration, equilibrium, reversible reaction, irreversible reaction
- ☐ Memory Work—This week, add the following elements to what you are working on memorizing:
  - ✓ 65-Tb-Terbium, 66-Dy-Dysprosium, 67-Ho-Holmium, 68-Er-Erbium
- ☐ Memory Work—Work on memorizing Le Chatelier's Principle – If a change is made to a reaction in equilibrium, the reaction will adjust itself to counter the effects of that change.

Sketch: Le Chatelier's Principle
- 🖾 Label the following: a reaction in equilibrium, the addition of more reactants causes the forward reaction to speed up, the addition of more products causes the reverse reaction to speed up

Writing
- ↝ Reading Assignment: *Usborne Illustrated Dictionary of Science* pp. 162-163 (Reversible Reactions)
- ↝ Additional Research Readings
  - 📖 Types of Reactions: *USE* pg. 78

Dates
- 🕐 1850-1936 – Henry Louis Le Chatelier lives. He is famous for the equilibrium law that he discovers during these years.

Schedules for Week 17
Two Days a Week

Day 1	Day 2
☐ Do the "Does the surface area affect cooking time?" experiment, then fill out the experiment sheet on SG pp. 126-127 ☐ Define concentration, equilibrium, reversible reaction, and irreversible reaction on SG pg. 103 ☐ Enter the dates onto the date sheets on SG pp. 8-13	☐ Read pp. 162-163 from *UIDS,* then discuss what was read ☐ Color and label the "Le Chatelier's Principle" sketch on SG pg. 125 ☐ Prepare an outline or narrative summary, write it on SG pp. 128-129

Supplies I Need for the Week:
- ✓ 2 potatoes
- ✓ Pot, Water
- ✓ Oven mitt
- ✓ Large Slotted Spoon

Things I Need to Prepare:

Five Days a Week

Day 1	Day 2	Day 3	Day 4	Day 5
☐ Do the "Does the surface area affect cooking time?" experiment, then fill out the experiment sheet on SG pp. 126-127 ☐ Enter the dates onto the date sheets on SG pp. 8-13	☐ Read pp. 162-163 from *UIDS,* then discuss what was read ☐ Write an outline on SG pg. 128	☐ Define concentration, equilibrium, reversible reaction, and irreversible reaction on SG pg. 103 ☐ Color and label the "Le Chatelier's Principle" sketch on SG pg. 125	☐ Read one or all of the additional reading assignments ☐ Write a report from what you learned on SG pg. 129	☐ Complete one of the Want More Activities listed **OR** ☐ Study a scientist from the field of Chemistry

Supplies I Need for the Week
- ✓ 2 potatoes
- ✓ Pot, Water
- ✓ Oven mitt
- ✓ Large Slotted Spoon

Things I Need to Prepare

Additional Information Week 17

Experiment Information

☞ **Introduction** – (*from the Student Guide*) The collision theory states that for a chemical reaction to occur the particles involved must hit one another with enough force or energy to break the bonds involved. In other words, there must collide with enough energy to overcome the activation energy required for the reaction to occur. If they don't, they will simply bounce off one another and no reaction will take place. Many factors such as temperature, concentration, and surface area can affect the rate at which the molecules collide in a reaction. In this experiment, you are going to examine how surface area can affect the rate of a reaction.

☞ **Results** – The students should see that the thinly sliced potato cooked the quickest, probably within 5 to 10 minutes. They should see that the quarter potato was the next to finish cooking, probably within 10 to 20 minutes. Then the half potato, finishing in 20 to 30 minutes. Finally, the whole potato should have taken the longest to fully cook.

☞ **Explanation** – When potatoes are cooked, the starch granules swell and burst, which causes the semi-crystalline structure to be broken up. The individual starch molecules then reconnect into a network in which water can be trapped. This leads to the softer texture of a cooked potato. The more overall surface area that the potato has exposed to the water, the quicker this reaction can occur, which explains why the thin slice of potato cooked much quicker than the other three potatoes.

☞ **Take if Further** – Cut several more thin slices of potatoes. Place half of them in a bowl with cold water and put the bowl in the fridge. Place the other half in another bowl with room temperature water and leave it out on the counter. After 4 hours check both bowls to see which potatoes are softer. Then, place both of the bowls in the microwave on high for five minutes to see what happens with the tenderness of the potato slices. How does an increase, or a decrease, in the temperature affect the reaction? (*You should see that the potato slice that sat on the counter was always the most tender.*)

Discussion Questions

1. What happens when a reaction reaches completion? (*UIDS pg. 162 - When a reaction reaches completion one or all of the reactants are used up and their products do not react together.*)

2. What is a reversible reaction? (*UIDS pg. 162 - A reversible reaction is a chemical reaction where the products can react together to form the original reactants.*)

3. What is the difference between a closed and open system? (*UIDS pg. 162 - In an open system, the chemicals in the reaction can escape. In a closed system, the chemicals in the reaction cannot escape or enter the reaction.*)

4. What happens when a reversible reaction reaches equilibrium? (*UIDS pg. 162 - When a reversible reaction reaches equilibrium, the forward and the backward reactions are occurring at the same speed.*)

5. What can affect the chemical equilibrium of a reaction? (*UIDS pg. 163 - Temperature, concentration, and pressure can all affect the equilibrium of a reaction.*)

6. What does Le Chatelier's principle say? (*UIDS pg. 163 - Le Chatelier's principle says that if changes are made to a system in equilibrium, the system adjusts itself to reduce the effects of the change.*)

Want More

☼ **Chemical Equations: Part 2** – Use the worksheet on pg. 263 of the Appendix to teach the students about balancing chemical equations for reversible reactions.
Answers to part 2 of the worksheet:

1. $2\,Al + 6\,HCl \longrightarrow 3\,H_2 + 2\,AlCl_3$
2. $N_2 + 2\,O_2 \rightleftharpoons 2\,NO_2$
3. $2\,NO_2 \rightleftharpoons N_2O_4$
4. $2\,SO_3 \rightleftharpoons 2\,SO_2 + O_2$
5. $N_2 + 3\,H_2 \rightleftharpoons 2\,NH_3$

Sketch Week 17

Le Chatelier's Principle

a reaction in equilibrium

the addition of more products causes the
reverse reaction to speed up

the addition of more reactants causes the
forward reaction to speed up

Student Assignment Sheet Week 18
Catalysts

Experiment: Can I speed up the reaction?
Materials:
- ✓ Carbonated water
- ✓ Sugar
- ✓ 2 cups

Procedure:
1. Read the introduction to this experiment and write down which reaction you think will go faster.
2. Label your two cups cup #1 and cup #2.
3. Pour one cup (240 mL) of carbonated water into each one of the cups. Wait till you see bubbles begin to form on the sides of the cup and record the time as your experiment start time.
4. Now, add 1 tsp (4 g) of sugar to cup #2. Record your observations about the reaction that is happening in both cups. When the reactions reach equilibrium (*i.e., no more bubbles are coming up*), record the end time for each cup in the experiment.
5. Draw conclusions and complete your experiment sheet.

Vocabulary & Memory Work
- ☐ Vocabulary: activation energy, catalyst
- ☐ Memory Work—This week, add the following elements to what you are working on memorizing:
  - ✓ 69-Tm-Thulium
  - ✓ 70-Yb-Ytterbium
  - ✓ 71-Lu-Lutetium
  - ✓ 72-Hf-Hafnium

Sketch: Reaction Pathway
- 🖾 Label the following: reactants, products, normal path of a reaction, activation energy, reaction path with the addition of a catalyst, activation energy when a catalyst has been introduced

Writing
- ꝰ Reading Assignment: *Usborne Illustrated Dictionary of Science* pp. 160-161 (Rates of Reactions)
- ꝰ Additional Research Readings:
  - 📖 Catalysts: *KSE* pg. 176, Enzymes: *KSE* pg. 177
  - 📖 Rates of Reactions: *USE* pg. 79

Dates
- 🕐 1767 – Joseph Priestley invents carbonated water when he discovers that he can infuse water with carbon dioxide by suspending a bowl of water above a beer.
- 🕐 1909 – Wilhelm Ostwald is awarded the Nobel Prize for Chemistry for his work with catalysts.

Schedules for Week 18
Two Days a Week

Day 1	Day 2
☐ Do the "Can I speed up the reaction?" experiment, then fill out the experiment sheet on SG pp. 132-133 ☐ Define activation energy and catalyst on SG pg. 103 ☐ Enter the dates onto the date sheets on SG pp. 8-13	☐ Read pp. 160-161 from *UIDS*, then discuss what was read ☐ Color and label the "Reaction Pathway" sketch on SG pg. 131 ☐ Prepare an outline or narrative summary, write it on SG pp. 134-135

Supplies I Need for the Week
✓ Carbonated water
✓ Sugar
✓ 2 cups

Things I Need to Prepare

Five Days a Week

Day 1	Day 2	Day 3	Day 4	Day 5
☐ Do the "Can I speed up the reaction?" experiment, then fill out the experiment sheet on SG pp. 132-133 ☐ Enter the dates onto the date sheets on SG pp. 8-13	☐ Read pp. 160-161 from *UIDS*, then discuss what was read ☐ Write an outline on SG pg. 134	☐ Define activation energy and catalyst on SG pg. 103 ☐ Color and label the "Reaction Pathway" sketch on SG pg. 131	☐ Read one or all of the additional reading assignments ☐ Write a report from what you learned on SG pg. 135	☐ Complete one of the Want More Activities listed **OR** ☐ Study a scientist from the field of Chemistry

Supplies I Need for the Week
✓ Carbonated water
✓ Sugar
✓ 2 cups

Things I Need to Prepare

120

Additional Information Week 18

Experiment Information

☞ **Introduction** – (*from the Student Guide*) The rate of a reaction can be changed by varying the temperature, pressure or concentration of the reactants involved. The rate can also be modified through the addition of a catalyst. A catalyst will alter the speed of a reaction, but it will not be changed by the process. In this experiment, you are going to look at how the addition of a catalyst can adjust the rate of a reaction.

☞ **Results** – The students should see that the reaction in cup #2 went much faster than the reaction in cup #1.

☞ **Explanation** – Carbonated water is made by injecting carbon dioxide into water under high pressure to create carbonic acid. This mixture is then stored in an air tight, pressurized bottle to prevent the reverse reaction from occurring. Here's what that reaction looks like:

$$CO_2(g) + H_2O(aq) \rightleftharpoons H_2CO_3(aq)$$

Once you open the bottle, the pressure decreases and the reverse reaction speeds up. Thus, we see carbon dioxide gas coming out of the solution as bubble. Once the reactions reach equilibrium again, you will see no more bubbles coming out of solution. The addition of sugar to the reaction causes the carbon dioxide to come out of the solution at a much quicker rate, but the sugar itself is not irreversibly changed in the process. It is merely dissolved into the solution, which makes sugar a catalyst to the above reaction.

☞ **Take if Further** – Test other materials in your kitchen, such as salt and baking soda, to see if they act as catalysts in the carbonated water reaction.

Discussion Questions

1. What is the rate of reaction? (*UIDS pg. 160 - The rate of a reaction is the speed at which the reactants are used up or the products are formed.*)
2. What is collision theory? (*UIDS pg. 160 - Collision theory explains why a change in the conditions under which a reaction occurs can affect the rate of the reaction. In other words, if more collisions occur, the rate of reaction is faster.*)
3. What factors can affect the rate of a reaction? (*UIDS pg. 161 - Heat, pressure, concentration, and surface area can all affect the rate of a reaction.*)
4. How can catalysts speed up a reaction? (*UIDS pg. 161 - Catalysts speed up a reaction by lowering the activation energy of the reaction.*)
5. What is a promoter? (*UIDS pg. 161 - A promoter increases the power of a catalyst, speeding the reaction up.*) An inhibitor? (*UIDS pg. 161 - An inhibitor reduces the power of a catalyst, slowing the reaction down.*)
6. What is an enzyme? (*UIDS pg. 161 - An enzyme is a catalyst found in a living thing.*)

Want More

✎ **Research Report** – Have the students research more about a specific catalyst or a process that uses catalysts, like the Haber process. Then, have them write a one to three paragraph report sharing what they have found.

Chemistry Unit 4: Chemical Reactions ~ Week 18 Catalysts

Sketch Week 18

The following sketch is not directly depicted in the text, but the information is shared. Your students may need additional help to label sketch.

Reaction Pathway

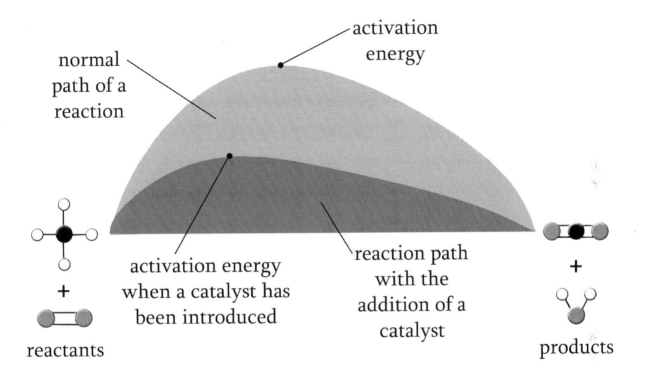

activation energy

normal path of a reaction

activation energy when a catalyst has been introduced

reaction path with the addition of a catalyst

reactants

products

Student Assignment Sheet Week 19
Oxidation and Reduction

Experiment: What happens when iron is oxidized?
  Materials:
  - ✓ Steel wool
  - ✓ Vinegar
  - ✓ 2 Jars with lids
  - ✓ Ammonia
  - ✓ Hydrogen peroxide

  Procedure:
  1. Read the introduction to this experiment and answer the question.
  2. The day before you do this experiment, prepare the iron acetate solution by placing the steel wool in one of the jars and covering it with vinegar. Put the lid on the jar and let the steel wool soak for 24 hours.
  3. The next day, put on your safety goggles before adding 2 TBSP (30 mL) of the iron acetate solution you made in the first jar to the second, empty jar. Then, add 2 to 4 TBSP (30 mL) of ammonia to the second jar. (*Basically, keep adding a tablespoon at a time until the color change you see remains for more than a brief moment.*) Record your observations.
  4. Next, add 1 TBSP (30 mL) of hydrogen peroxide to the second jar. (*Again, you may need to add an additional tablespoon of hydrogen peroxide so that the color remains.*) Record your observations.
  5. Draw conclusions and complete your experiment sheet.

Vocabulary & Memory Work
  ☐ Vocabulary: oxidation, reduction, redox reaction
  ☐ Memory Work—This week, add the following elements to what you are working on memorizing:
  - ✓ 73-Ta-Tantalum, 74-W-Tungsten, 75-Re-Rhenium, 76-Os-Osmium

Sketch: Oxidation/Reduction Reaction
  ▨ Label the following: the oxidizing agent gives oxygen to another molecule, the reducing agent gives hydrogen to another molecule, the reduced molecule has gained a hydrogen atom, the oxidized molecule has gained an oxygen atom

Writing
  ᕦ Reading Assignment: *Usborne Illustrated Dictionary of Science* pp. 148-149 (Oxidation and Reduction)
  ᕦ Additional Research Readings:
  - 📖 Oxidation and Reduction: *KSE* pg. 178
  - 📖 Oxidation and Reduction: *USE* pp. 80-81

Dates
  🕐 No dates to be entered this week.

  Chemistry Unit 4: Chemical Reactions ~ Week 19 Oxidation and Reduction

123

Schedules for Week 19
Two Days a Week

Day 1	Day 2
☐ Do the "What happens when iron is oxidized?" experiment, then fill out the experiment sheet on SG pp. 138-139 ☐ Define oxidation, reduction, and redox reaction on SG pp. 103-104 ☐ Enter the dates onto the date sheets on SG pp. 8-13	☐ Read pp. 148-149 from *UIDS,* then discuss what was read ☐ Label the "Oxidation/Reduction Reaction" sketch on SG pg. 137 ☐ Prepare an outline or narrative summary, write it on SG pp. 140-141

Supplies I Need for the Week
- ✓ Steel wool, Vinegar
- ✓ Jar with lid
- ✓ Ammonia, Hydrogen peroxide

Things I Need to Prepare

Five Days a Week

Day 1	Day 2	Day 3	Day 4	Day 5
☐ Do the "What happens when iron is oxidized?" experiment, then fill out the experiment sheet on SG pp. 138-139 ☐ Enter the dates onto the date sheets on SG pp. 8-13	☐ Read pp. 148-149 from *UIDS,* then discuss what was read ☐ Write an outline on SG pg. 140	☐ Define oxidation, reduction, and redox reaction on SG pp. 103-104 ☐ Label the "Oxidation/Reduction Reaction" sketch on SG pg. 137	☐ Read one or all of the additional reading assignments ☐ Write a report from what you learned on SG pg. 141	☐ Complete one of the Want More Activities listed **OR** ☐ Take the Unit 4 Test

Supplies I Need for the Week
- ✓ Steel wool, Vinegar
- ✓ Jar with lid
- ✓ Ammonia, Hydrogen peroxide

Things I Need to Prepare

Chemistry Unit 4: Chemical Reactions ~ Week 19 Oxidation and Reduction

Additional Information Week 19

Note

🖋 **Redox Pneumonic** – To help your students memorize what happens in a redox reaction, you can use the pneumonic: OIL RIG, which stands for: Oxidation is lost, reduction is gained.

Experiment Information

☞ **Introduction** – (*from the Student Guide*) Combustion, rusting and respiration are all examples of oxidation and reduction reactions. These reactions typically involve the movement of oxygen and hydrogen atoms, however an oxidation/reduction reaction can occur even when oxygen or hydrogen are not present. The key is to have the movement of electrons present in the reaction. In other words, an atom is oxidized when it loses an electron and reduced when it gains one. In this experiment, you are going to look at what happens to iron atoms when they are oxidized.

☞ **Results** – The iron acetate solution should start out relatively clear. After ammonia is added, the solution should have a greenish tint. After the hydrogen peroxide is added, the solution should have a reddish, rust-colored tint.

☞ **Explanation** – The iron in the reaction is being oxidized, or rather it is losing electrons in the reaction. Iron is a metal that has two ions, each of which have different colors, which makes them easy to determine. Ferrous iron, which is iron that has given away two of its electrons is greenish tint, while ferric iron, which has given away three of its electrons is rust colored. So the redox reaction that occurred in step 4 of the experiment would be:

Ferrous iron + hydrogen peroxide ⟶ ferric iron + water + oxygen

$$3Fe_2^+ + 2H_2O_2 \longrightarrow 2Fe^{3+} + 2H_2O + O_2$$

☞ **Take if Further** – Watch another redox reaction by taking the steel wool out of the vinegar in the first jar and setting the steel wool on the jar lid. After thirty minutes check to see what has happened to the steel wool. (*It should begin to rust, or oxidize.*)

Discussion Questions

1. What do oxidation and reduction refer to? (*UIDS pg. 148 - Oxidation and reduction refer to the gain and loss of either oxygen, hydrogen, or electrons by a substance in a chemical reaction.*)

2. What happens to a substance when it is oxidized? (*UIDS pg. 148 - When a substance is oxidized, it gains oxygen, loses hydrogen, or loses electrons in a chemical reaction.*) When it is reduced? (*UIDS pg. 148 - When a substance is reduced, it loses oxygen, gains hydrogen, or gains electrons in a chemical reaction.*)

3. What happens to the oxidizing agent? (*UIDS pp. 148 & 149 - The oxidizing agent is always reduced in a reaction. This substance readily gains electrons.*) The reducing agent? (*UIDS pg. 148 - The reducing agent is always oxidized in a reaction. This substance readily gains electrons.*)

4. Why do oxidation and reduction reactions always occur together? (*UIDS pg. 148 - Oxidation and reduction must occur together because one substance, the oxidizing agent, is reduced during oxidation, while the other substance, the reducing agent, is oxidized during*

reduction.)

5. What is the oxidation state of an element? (*UIDS pg. 149 - The oxidation state of an element is number of electrons that have been added to or removed from an atom when it forms a compound.*)

Want More

☞ **Chemical Equations: Part 3** – Use the worksheet on pg. 264 of the Appendix to review balancing chemical equations with the students.

Answers to part 3 of the worksheet:

✓ $FeO + PdF_2 \longrightarrow FeF_2 + PdO$
✓ $Si(OH)_4 + 4\ NaBr \longrightarrow SiBr_4 + 4\ NaOH$
✓ $2\ RbNO_3 + BeF_2 \longrightarrow Be(NO_3)_2 + 2\ RbF$
✓ $N_2 + 3\ F_2 \longrightarrow 2\ NF_3$
✓ $H_2 + I_2 \rightleftharpoons 2\ HI$

Sketch Week 19

The following sketch is not directly depicted in the text, but the information is shared. Your students may need additional help to label sketch.

Oxidation/Reduction Reaction

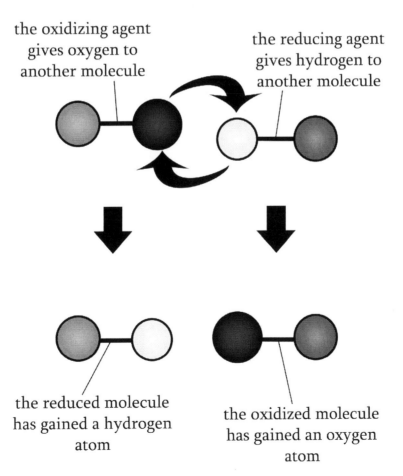

the oxidizing agent gives oxygen to another molecule

the reducing agent gives hydrogen to another molecule

the reduced molecule has gained a hydrogen atom

the oxidized molecule has gained an oxygen atom

Chemistry Unit 4: Chemical Reactions
Unit Test Answers

Vocabulary Matching

1. B
2. G
3. A
4. C
5. E
6. I
7. D
8. L
9. F
10. K
11. O
12. J
13. Q
14. R
15. H
16. T
17. N
18. S
19. P
20. M

True or False

1. True
2. False (*Chemical bonding always involves the outermost electrons of an atom.*)
3. False (*The substances present at the beginning of the reaction are the reactants. The substances present at the end of the reaction are the products.*)
4. True
5. True
6. False (*The reactivity of an element depends on its ability to lose or gain an electron.*)
7. True
8. True
9. False (*A promoter increases the power of a catalyst, speeding the reaction up. An inhibitor reduces the power of a catalyst, slowing the reaction down.*)
10. False (*Catalysts can speed up a reaction by decreasing the activation energy of the particular reaction.*)
11. False (*The oxidizing agent is always reduced in a reaction. The reducing agent is always oxidized in a reaction.*)
12. True

Short Answer

1. Ionic bonding is a strong chemical bond that is caused by the attraction between two ions of opposite charges. Covalent bonding is a chemical bond between two atoms, in which each atom shares an electron. Metallic bonding is a chemical bond where positive metal ions create a lattice structure with freely moving electrons between them.
2. In an endothermic reaction the heat energy needed to break the bonds of the reactants is greater than the heat energy given off when the new bonds of the products are formed. In an exothermic reaction the heat energy needed to break the bonds of the reactants is less than the heat energy given off when the new bonds of the products are formed.
3. 6.02×10^{23}, Avogadro's number tells us the number of particles per mole.
4. The reactivity series tell us how reactive certain metals are. The metals at the top are very reactive, while those at the bottom are the least reactive.
5. Le Chatelier's principle says that if changes are made to a system in equilibrium, the system adjusts itself to reduce the effects of the change.

6. Collision theory explains why a change in the conditions under which a reaction occurs can affect the rate of the reaction. In other words, if more collisions occur, the faster the rate of reaction.

7. When a substance is oxidized, it gains oxygen, loses hydrogen, or loses electrons in a chemical reaction. When a substance is reduced, it loses oxygen, gains hydrogen, or gains electrons in a chemical reaction.

8. The law of conservation of mass says that matter cannot be created or destroyed during a chemical reaction.

9. 53-I-Iodine, 54-Xe-Xeon, 55-Cs-Cesium, 56-Ba-Barium, 57-La-Lanthanum, 58-Ce-Cerium, 59-Pr-Praseodymium, 60-Nd-Neodymium, 61-Pm-Promethium, 62-Sm-Samarium, 63-Eu-Europium, 64-Gd-Gadolinium, 65-Tb-Terbium, 66-Dy-Dysprosium, 67-Ho-Holmium, 68-Er-Erbium, 69-Tm-Thulium, 70-Yb-Ytterbium, 71-Lu-Lutelium, 72-Hf-Hafnium, 73-Ta-Tantalum, 74-W-Tungsten, 75-Re-Rhenium, 76-Os-Osmium

Chemistry Unit 4: Chemical Reactions
Unit Test

Vocabulary Matching:

1. Ion ___

2. Ionic Bonding ___

3. Covalent Bonding ___

4. Metallic Bonding ___

5. Chemical Reaction ___

6. Endothermic Reaction ___

7. Exothermic Reaction ___

8. Product ___

9. Reactant ___

10. Moles ___

11. Reactivity ___

12. Concentration ___

13. Equilibrium ___

14. Irreversible Reaction ___

15. Reversible Reaction ___

16. Activation Energy ___

17. Catalyst ___

A. A chemical bond between two atoms, in which each atom shares an electron.

B. An atom or group or atoms that has become charged by gaining or losing one or more electrons.

C. A chemical bond where positive metal ions form a lattice structure with freely moving electrons between them.

D. A chemical reaction that gives off heat.

E. An interaction between the atoms of two substances where the atoms rearrange to form two new molecules.

F. The substance present at the beginning of a chemical reaction.

G. A strong chemical bond that is formed by the attraction between two ions of opposite charges.

H. A reaction in which, under the right conditions, the products can be turned back into the reactants.

I. A chemical reaction that takes in heat.

J. A measure of the strength of a solution, i.e., the amount of the solute in the solvent.

K. The SI unit for the amount of a substance.

L. A new substance produced in a chemical reaction.

M. A reaction that primarily involves the transfer of electrons between two substances.

N. A substance which speeds up a chemical reaction without being changed by the reaction.

O. The tendency of a substance to react with other substances.

P. Occurs when an atom gains electrons in a reaction, or when a substance gains a hydrogen or loses an oxygen atom in a chemical reaction.

Q. The point at which the rate of the forward reaction is equal to the rate of the backward reaction.

18. Oxidation ____

R. A reaction in which the products can never be turned back into the reactants.

19. Reduction ____

S. Occurs when an atom loses electrons in a reaction, or when a substance loses a hydrogen or gains an oxygen atom in a chemical reaction.

20. Redox Reaction ____

T. The minimum amount of energy a reaction needs to start.

True or False

1. _____ A cation has a positive charge. An anion has a negative charge.

2. _____ Chemical bonding always involves the innermost electrons of an atom.

3. _____ The substances present at the beginning of the reaction are the products. The substances present at the end of the reaction are the reactants.

4. _____ In a chemical reaction, bonds are broken and made, which requires energy.

5. _____ Chemical equations tell all scientists what substances are involved in a reaction and in what proportions they are needed.

6. _____ The reactivity of an element depends on its ability to lose or gain a proton.

7. _____ An increase in temperature, pressure, or concentration can all affect the equilibrium of a reaction.

8. _____ A reversible reaction is a chemical reaction where the products can react together to form the original reactants.

9. _____ An inhibitor increases the power of a catalyst, speeding the reaction up. A promoter reduces the power of a catalyst, slowing the reaction down.

10. _____ Catalysts can speed up a reaction by increasing the activation energy of the particular reaction.

11. _____ The reducing agent is always reduced in a reaction. The oxidizing agent is always oxidized in a reaction.

12. _____ Oxidation and reduction reactions always occur together.

Short Answer

1. Briefly explain ionic bonding, covalent bonding, and metallic bonding.

2. What is the difference between exothermic and endothermic reactions?

3. What is Avogadro's number, and what does it tell us?

4. What does the reactivity series tell us?

5. What does Le Chatelier's principle say?

6. What is the collision theory?

7. What happens to a substance when it is reduced? What happens to a substance when it is oxidized?

8. What is the law of conservation of mass?

9. Write elements 53-76 in order along with their abbreviation from the periodic table.

- Lutetium
- Hafnium
- Iodine
- Osmium
- Cesium
- Gadolinium
- Barium
- Lanthanum

- Cerium
- Promethium
- Tungsten
- Holmium
- Erbium
- Europium
- Terbium
- Dysprosium

- Thulium
- Ytterbium
- Samarium
- Tantalum
- Rhenium
- Praseodymium
- Xenon
- Neodymium

53. _____

54. _____

55. _____

56. _____

57. _____

58. _____

59. _____

60. _____

61. _____

62. _____

63. _____

64. _____

65. _____

66. _____

67. _____

68. _____

69. _____

70. _____

71. _____

72. _____

73. _____

74. _____

75. _____

76. _____

Chemistry: Unit 5

Acids and Bases

Unit 5: Acids and Bases
Overview of Study

Sequence of Study

Week 20: Acids
Week 21: Bases
Week 22: Measuring Acidity (pH)
Week 23: Neutralization and Salts

Materials by Week

Week	Materials
20	Cranberry juice, Lemon juice, Baking soda, Clear cup
21	6 cups, Red cabbage solution, Water, Vinegar, Baking soda, Sprite, Ammonia, Lemon Juice, Eye dropper
22	Lemon, Tomato, Saliva, Milk, Bleach, Toothpaste, Dish Soap, pH paper, Gloves
23	Vinegar, Ammonia, Red cabbage solution, Water, Safety glasses

Vocabulary for the Unit

1. **Acid** – A hydrogen containing compound that splits in water to give hydrogen ions.
2. **Acidic Solution** – A solution that contains an acid, or a solution that has a pH less than seven.
3. **Dissociation** – The process by which a substance is split up into its ions in a solution.
4. **Alkali** – A base that dissolves in water to form an hydroxide ion.
5. **Alkaline Solution** – A solution that contains a base, or a solution that has a pH greater than seven.
6. **Base** – A compound that reacts with an acid to produce water and a salt.
7. **Indicator** – A substance that changes color as the pH of the solution changes.
8. **Buffer** – A solution that resists changes in pH.
9. **pH** – A measure of the acidity or alkalinity of a solution.
10. **Titration** – A method of finding the concentration of acidic or alkaline solutions.
11. **Neutralization** – The process by which you make a solution neither acidic nor alkaline.
12. **Salt** – A type of compound that is formed when an acid and a base react.

Memory Work for the Unit

The Elements of the Periodic Table – The following elements will be memorized in this unit:
  ✓ 77-Ir-Iridium

- ✓ 78-Pt-Platinum
- ✓ 79-Au-Gold
- ✓ 80-Hg-Mercury
- ✓ 81-Tl-Thallium
- ✓ 82-Pb-Lead
- ✓ 83-Bi-Bismuth
- ✓ 84-Po-Polonium
- ✓ 85-At-Astatine
- ✓ 86-Rn-Radon
- ✓ 87-Fr-Francium
- ✓ 88-Ra-Radium
- ✓ 89-Ac-Actinium
- ✓ 90-Th-Thorium
- ✓ 91-Pa-Protactinium
- ✓ 92-U-Uranium

Notes

Student Assignment Sheet Week 20
Acids

Experiment: Can I change an acid?
Materials:
- ✓ Cranberry juice
- ✓ Lemon juice
- ✓ Baking soda
- ✓ Clear cup

Procedure:
1. Read the introduction to this experiment and then write down what you think will happen when you add the baking soda and lemon juice to the cranberry juice.
2. Begin by pouring 1 cup (240 mL) of cranberry juice into the cup. Then, add 1 TBSP (15 mL) of lemon juice. Note any changes and record the observations.
3. Then, add 1 TBSP (14 g) of baking soda. Note any changes and record the observations.
4. Finally, add another TBSP (15 mL) of lemon juice. Note any changes and record the observations.
5. Draw conclusions and complete the experiment sheet.

Vocabulary & Memory Work
- ☐ Vocabulary: acid, acidic solution, dissociation
- ☐ Memory Work—This week, add the following elements to what you are working on memorizing:
  - ✓ 77-Ir-Iridium, 78-Pt-Platinum, 79-Au-Gold, 80-Hg-Mercury

Sketch Assignment: Acids
- 🖾 Label the following: dissolved in water, hydrogen chloride gas, hydrogen ion, chloride ion

Writing
- ᕤ Reading Assignment: *Usborne Science Encyclopedia* pg. 84 (Acids) and *Usborne Illustrated Dictionary of Science* pg. 150 (Acids)
- ᕤ Additional Research Readings:
  - 📖 Acids: *KSE* pg. 184

Dates
- 🕐 11th century – Arabic chemists discover how to make acidic compounds such as sulfuric, nitric and hydrochloric acids.
- 🕐 1675 – Robert Boyle, an Irish chemist, suggests that acids might be made up of sharp pointed particles.
- 🕐 1770's – Antoine Lavoisier, a French chemist, suggests that acids form oxygen compounds when dissolved in water, so they must all contain oxygen.
- 🕐 1887 – Svante Arrhenius, a Swedish chemist, develops the modern theory on acids. He states that all acids contain hydrogen ions, which give these compounds their unique properties.

Schedules for Week 20
Two Days a Week

Day 1	Day 2
☐ Do the "Can I change an acid?" experiment, then fill out the experiment sheet on SG pp. 148-149 ☐ Define acid, acidic solution, and dissociation on SG pg. 144 ☐ Enter the dates onto the date sheets on SG pp. 8-13	☐ Read pg. 84 from *USE* and pg. 150 from *UIDS,* then discuss what was read ☐ Color and label the "Strong Acids vs. Weak Acids" sketch on SG pg. 147 ☐ Prepare an outline or narrative summary, write it on SG pp. 150-151

Supplies I Need for the Week
- ✓ Cranberry juice
- ✓ Lemon juice
- ✓ Baking soda
- ✓ Clear cup

Things I Need to Prepare

Five Days a Week

Day 1	Day 2	Day 3	Day 4	Day 5
☐ Do the "Can I change an acid?" experiment, then fill out the experiment sheet on SG pp. 148-149 ☐ Enter the dates onto the date sheets on SG pp. 8-13	☐ Read pg. 84 from *USE* and pg. 150 from *UIDS,* then discuss what was read ☐ Write an outline on SG pg. 150	☐ Define acid, acidic solution, and dissociation on SG pg. 144 ☐ Color and label the "Strong Acids vs. Weak Acids" sketch on SG pg. 147	☐ Read one or all of the additional reading assignments ☐ Write a report from what you learned on SG pg. 151	☐ Complete one of the Want More Activities listed **OR** ☐ Study a scientist from the field of Chemistry

Supplies I Need for the Week
- ✓ Cranberry juice
- ✓ Lemon juice
- ✓ Baking soda
- ✓ Clear cup

Things I Need to Prepare

Chemistry Unit 5: Acids and Bases ~ Week 20: Acids

Additional Information Week 20

Experiment Information

☞ **Introduction** – (*from the Student Guide*) Acids release positive hydrogen ions when in a solution. They are in batteries, in our stomachs and in the food we eat and drink. Acids are responsible for the sour taste that you find in food. Some of the fruits we eat, such as pears and blueberries, contain pigments that change colors in the presence of an acid or base. The pigment anthocyanin, which is found in cranberries, is red when in the presence of an acid and blue in the presence of a base. In this experiment, you are going to try to alter the color of acidic cranberry juice with other common kitchen substances.

☞ **Results** – The students should see that when they added the lemon juice, there was no change in the color. They should see the juice turn purple or blue after adding the of baking soda. Finally, they should see the solution turn red again with the addition of the second amount of lemon juice.

☞ **Explanation** – Remember that cranberry juice contains the pigment anthocyanin, which is red in the presence of an acid and blue in the presence of a base. Lemon juice is a weak acid, so when you add it to the acidic cranberry juice, there is no change. On the other hand, baking soda is a weak base. When you add it to the acidic cranberry solution, it begins to neutralize the acid, turning the solution purple or blue. When you add more lemon juice, the solution turns acidic, or red, once again. (**Note**—*You may have also seen some fizzing when you added the baking soda to the cranberry solution. This fizz is from bubbles of carbon dioxide gas which form when baking soda is added to an acid.*)

☞ **Take it Further** – Pour a fresh cup of cranberry juice and try out several other kitchen substances, such as water, ammonia, or vinegar, to see if they alter the color of the juice. (*You should see that the water and vinegar do not affect the color of the juice, while the ammonia turns it purple or blue. This is because vinegar is an acid, water is neutral and ammonia is a base.*)

Discussion Questions

1. What is unique to acids? (*USE pg. 84 - Acids form hydrogen ions in solution, which gives them their unique properties, but only when they are dissolved.*)
2. What is the difference between strong acids and weak acids? (*USE pg. 84 - Strong acids completely split up in water to form a large number of hydrogen ions, while weak acids only contain a few hydrogen ions.*)
3. What are organic acids? (*USE pg. 84 - Organic acids are acidic compounds produced by living things.*)
4. How do acids behave? (*USE pg. 84 - Acids react with metals to form salts and hydrogen gas. Acids also react with carbonates to give salt, carbon dioxide, and water.*)
5. What happens to acids in water? (*UIDS pg. 150 - When acids dissolve in water they yield positively charged hydrogen ions, which combine with water to form hydroxonium ions.*)
6. What do acids always do to carbonates? (*UIDS pg. 150 - Acids always break down carbonates to form carbon dioxide.*)

Want More

🗐 **Acid/Base Poster** – This project will be done over two weeks. The students will take a piece of poster board and divide it into two sections. On one side they will write "Acids" and on the other side they will write "Bases". For this week, have them research about common acids that can be found in the house. Then, have the students fill the acids' side of the poster with the information they have learned. They can choose to draw pictures, cut some out of a magazine, or simply list the household items that contain acids. Either way they choose to depict them, they should include the common name, the chemical name for the acid it contains and how the household item can be dangerous.

Sketch Week 20

Acids

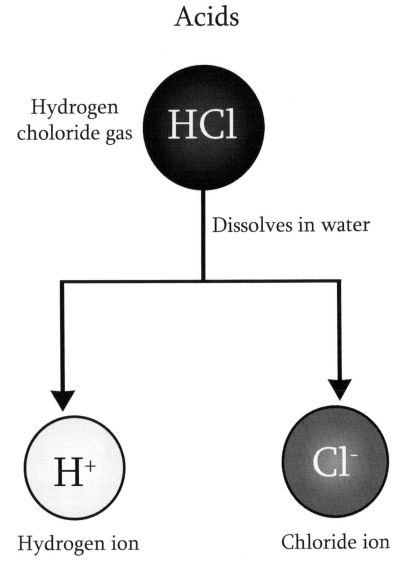

Hydrogen choloride gas — HCl

Dissolves in water

H+ — Hydrogen ion

Cl- — Chloride ion

Student Assignment Sheet Week 21
Bases

Experiment: Is it an acid or a base?

Materials:
- ✓ 6 cups
- ✓ Red cabbage solution
- ✓ Water
- ✓ Vinegar
- ✓ Baking soda
- ✓ Sprite
- ✓ Ammonia
- ✓ Lemon Juice
- ✓ Eye dropper

Procedure:

How to make the red cabbage solution – The night before you do this experiment you will need to make the red cabbage solution. Have your parent or teacher cut up a head of cabbage, place it in a pot and put it on the stove. Turn the heat to high and boil the cabbage for 15 minutes. Remove the cooked cabbage and let the solution cool. Once it is cool put it in a container and store it in a glass jar in the refrigerator. You will also need this solution for the experiment in week 23.

1. Read the introduction to this experiment and write down whether you think vinegar, baking soda, sprite, ammonia and lemon juice are acids or bases.
2. Label the cups #1 through #6. Pour ¼ cup (60 mL) of water into cup #1, a ¼ cup (60 mL) of vinegar into cup #2, ¼ cup (60 mL) of Sprite into cup #3, ¼ cup (60 mL) of ammonia into cup #5, and ¼ cup (60 mL) of lemon juice into cup #6. Add ¼ cup (60 mL) of water to cup #4 and dissolve 1 TBSP (14 g) of baking soda in it.
3. Add 1 tsp (5 mL) of red cabbage juice to each cup, mix thoroughly. Be sure to either thoroughly rinse the spoon after each one or use a different spoon for each cup.
4. Compare the color of the solution in each cup to the color of the original solution. If the solution is the same color as the original, the substance is neutral. If the solution turned to a light red-purple, the substance was a weak acid. If the solution was red, the substance was a strong acid. If the solution turned blue, the substance was a weak base. If the solution was blue-green, the substance was a strong base.
5. Record the color of the solution on the results chart of the experiment sheet and then decide if the solutions were acidic, neutral or basic.
6. Draw conclusions and complete the experiment sheet.

Vocabulary & Memory Work

- ☐ Vocabulary: alkali, alkaline solution, base, indicator
- ☐ Memory Work—This week, add the following elements to what you are working on memorizing:
  - ✓ 81-Tl-Thallium, 82-Pb-Lead, 83-Bi-Bismuth, 84-Po-Polonium

Sketch: Bases

- 🖼 Label the following: dissolved in water, sodium hydroxide chloride gas, sodium ion, hydroxide ion

Writing

- ᷂ Reading Assignment: *Usborne Science Encyclopedia* pg. 85 (Bases) and *Usborne Illustrated Dictionary of Science* pg. 151 (Bases)
- ᷂ Additional Research Readings: Bases: *KSE* pg. 185

Dates

- 🕐 1754 – French chemist Guillaume-François Rouelle states that a base is a substance that reacts with an acid to produce a solid (or salt).

Schedules for Week 21
Two Days a Week

Day 1	Day 2
☐ Do the "Is it an acid or a base?" experiment, then fill out the experiment sheet on SG pp. 154-155 ☐ Define alkali, alkaline solution, base, and indicator on SG pg. 144 ☐ Enter the dates onto the date sheets on SG pp. 8-13	☐ Read pg. 85 from *USE* and pg. 151 from *UIDS,* then discuss what was read ☐ Color and label the "Strong Bases vs. Weak Bases" sketch on SG pg. 153 ☐ Prepare an outline or narrative summary, write it on SG pp. 156-157

Supplies I Need for the Week
✓ 6 cups, Red cabbage solution, Water
✓ Vinegar, Baking soda, Sprite, Ammonia
✓ Lemon Juice, Eye dropper

Things I Need to Prepare

Five Days a Week

Day 1	Day 2	Day 3	Day 4	Day 5
☐ Do the "Is it an acid or a base?" experiment, then fill out the experiment sheet on SG pp. 154-155 ☐ Enter the dates onto the date sheets on SG pp. 8-13	☐ Read pg. 85 from *USE* and pg. 151 from *UIDS,* then discuss what was read ☐ Write an outline on SG pg. 156	☐ Define alkali, alkaline solution, base, and indicator on SG pg. 144 ☐ Color and label the "Strong Bases vs. Weak Bases" sketch on SG pg. 153	☐ Read one or all of the additional reading assignments ☐ Write a report from what you learned on SG pg. 157	☐ Complete one of the Want More Activities listed **OR** ☐ Study a scientist from the field of Chemistry

Supplies I Need for the Week
✓ 6 cups, Red cabbage solution, Water
✓ Vinegar, Baking soda, Sprite, Ammonia
✓ Lemon Juice, Eye dropper

Things I Need to Prepare

144

Additional Information Week 21

Notes

❧ **A Point of Clarification** – The terms "alkali" and "base" are sometimes used interchangeably, but there is a slight difference between the two. Bases are substances that react with acids to form a salt and water, such as calcium carbonate. Alkalis are substance that when dissolved in water form a hydroxide ion, such as sodium hydroxide. In other words, all alkalis are bases, but not all bases are alkalis.

Experiment Information

☞ **Introduction** – (*from the Student Guide*) The solutions you see in the everyday life all have either acidic or basic properties. In other words, they can either release positive hydrogen ions into a solution (acid) or they can release negative hydroxide ions into a solution (base). In the laboratory, scientists can use indicators to tell whether or not a solution is acidic or basic. This indicator will change color depending upon the properties of the solution. In this experiment, you are going use an indicator to look at several different common household substances to determine whether they are acids or bases.

☞ **Results** – The students should have the following results chart:

Substance	Color	Acid or Base
Water	purple-blue	neutral
Vinegar	pink	acid
Baking soda	blue	base
Sprite	pink	acid
Ammonia	blue-green	base
Lemon juice	red	acid

☞ **Explanation** – The red cabbage acts as an indicator, meaning that it changes color in the presence of an acid or a base. A typical kitchen is full of weak acids, such as vinegar, and weak bases, such as baking soda. We can also readily find strong acids, such as lemon juice and strong bases, such as bleach. Although there are many safe acids and bases in the kitchen, as a general rule you should avoid touching these chemicals without the proper protection.

☞ **Take it Further** – Try other household chemicals, like toothpaste (*base*), powdered detergent (*base*), milk (*acid*), or apple juice (*acid*).

Discussion Questions

1. What are bases? (*USE pg. 85 - Bases are the opposite of acids and react with acids to form water and salt.*)
2. What are some uses of bases? (*USE pg. 85 - Bases can be used to dissolve dirt for cleaning or resin in paper making.*)
3. What do bases do to acids? (*UIDS pg. 151 - Bases neutralize acids by accepting hydrogen ions.*)
4. What do all alkalis have in common? (*UIDS pg. 151 - All alkalis dissolve in water to form hydroxide ions, forming alkaline solutions.*)

Chemistry Unit 5: Acids and Bases ~ Week 21: Bases

5. What is neutralization? (*UIDS pg. 151 - Neutralization is a process that occurs when an acid and a base meet and cancel one another out to produce salt and water.*)

Want More

○ **Acid/Base Poster** – This is the second week for this project. The students have taken a piece of poster board and divided it into two sections and they have completed the acid side of their poster already. For this week, have them research about common bases that can be found in the house. Then, have the students fill the bases side of the poster with the information they have learned. They can choose to draw pictures, cut some out of a magazine, or simply list the household items that contain bases. Either way they choose to depict them, they should include the common name, the chemical name for the base it contains and how the household item could be dangerous.

Sketch Week 21

Bases

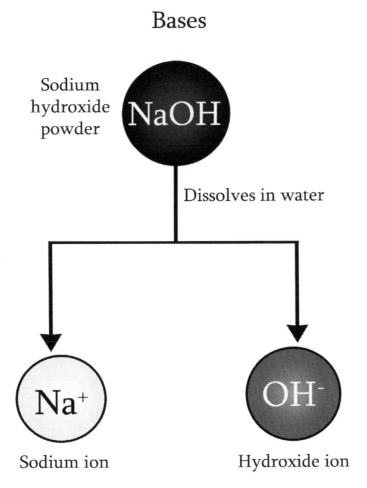

Sodium hydroxide powder

NaOH

Dissolves in water

Na⁺

Sodium ion

OH⁻

Hydroxide ion

Student Assignment Sheet Week 22
Measuring Acidity (pH)

Experiment: Measuring pH

Materials:

- ✓ Lemon
- ✓ Tomato
- ✓ Saliva
- ✓ Milk
- ✓ Bleach
- ✓ Toothpaste
- ✓ Dish soap
- ✓ pH paper
- ✓ Gloves

> **pH Scale**
> ☞ pH 0 to just under 7 is an acid.
> ☞ pH 7 is neutral.
> ☞ pH just over 7 to 14 is a base.

Procedure:

1. Read the introduction to this experiment.
2. Put on a pair of gloves and lay out each of your test substances in a row. Touch the pH paper to the surface of the test substance and wait for the paper to change color.
3. Compare the color of the paper with the scale included with your pH paper and record the approximate pH of the substance. Then, use the number scale above to determine if the substance is an acid, base or neutral.
4. Draw conclusions and complete the experiment sheet.

Vocabulary & Memory Work

- ☐ Vocabulary: buffer, pH
- ☐ Memory Work—This week, add the following elements to what you are working on memorizing:
  - ✓ 85-At-Astatine, 86-Rn-Radon, 87-Fr-Francium, 88-Ra-Radium

Sketch: pH scale

- ▦ Label the following: the pH scale from 0 to 14, acidic range, neutral, basic range
- ▦ Then, use varying shades of red to show that the strength of an acid increases as you go down the scale and varying shades of blue to show that the strength of a base increases as you go up.

Writing

- ↬ Reading Assignment: *Usborne Science Encyclopedia* pg. 86 (pH and Indicators) and *Usborne Illustrated Dictionary of Science* pg. 152 (Strength and Concentration)
- ↬ Additional Research Readings:
  - 📖 Indicators and pH: *KSE* pg. 186

Dates

- 🕐 1909 – Danish chemist Soren Sorenson comes up with a logarithmic scale to show the concentration of the hydrogen ions in a solution, which ranged from 0 to 14. Today this scale is known as the pH scale.

Schedules for Week 22
Two Days a Week

Day 1	Day 2
☐ Do the "Measuring pH" experiment, then fill out the experiment sheet on SG pp. 160-161 ☐ Define buffer and pH on SG pg. 144-145 ☐ Enter the dates onto the date sheets on SG pp. 8-13	☐ Read pg. 86 from *USE* and pg. 152 from *UIDS,* then discuss what was read ☐ Color and label the "pH Scale" sketch on SG pg. 159 ☐ Prepare an outline or narrative summary, write it on SG pp. 162-163

Supplies I Need for the Week
✓ Lemon, Tomato, Saliva
✓ Milk, Bleach, Toothpaste, Dish soap
✓ pH paper, gloves

Things I Need to Prepare

Five Days a Week

Day 1	Day 2	Day 3	Day 4	Day 5
☐ Do the "Measuring pH" experiment, then fill out the experiment sheet on SG pp. 160-161 ☐ Enter the dates onto the date sheets on SG pp. 8-13	☐ Read pg. 86 from *USE* and pg. 152 from *UIDS,* then discuss what was read ☐ Write an outline on SG pg. 162	☐ Define buffer and pH on SG pg. 144-145 ☐ Color and label the "pH Scale" sketch on SG pg. 159	☐ Read one or all of the additional reading assignments ☐ Write a report from what you learned on SG pg. 163	☐ Complete one of the Want More Activities listed **OR** ☐ Study a scientist from the field of Chemistry

Supplies I Need for the Week
✓ Lemon, Tomato, Saliva
✓ Milk, Bleach, Toothpaste, Dish soap
✓ pH paper, gloves

Things I Need to Prepare

Chemistry Unit 5: Acids and Bases ~ Week 22: Measuring Acidity (pH)

Additional Information Week 22

Experiment Information

☞ **Introduction** – (*from the Student Guide*) The pH of a solution is a measure of the acidity or alkalinity of the solution. The pH scale goes from zero to fourteen. A solution with a pH below seven is considered an acid and a solution with a pH above seven is considered a base. If the solution has a pH of seven, it is considered to be neutral. The closer an acidic solution is to having a pH of zero, the stronger it is, and the stronger an alkaline solution is to having a pH of fourteen the stronger it is. In this experiment, you are going to objectively measure the pH of several common household substances.

☞ **Results** – The students should have the following results in his chart:

Substance	pH	Acid, Base or Neutral
Lemon	2	Acid
Tomato	4	Acid
Saliva	6	Acid
Milk	6	Acid
Bleach	13	Base
Toothpaste	9	Base
Liquid dish soap	8	Base

☞ **Explanation** – The purpose of this experiment was to have the students gain familiarity with using pH paper. If they were able to successfully complete their results chart, they have reached the goal of the exercise.

☞ **Take it Further** – Have the students choose other substances to test with their pH paper.

Discussion Questions

1. What does pH stand for, and what does it measure? (*USE pg. 86 - pH stands for "power of hydrogen" and it measures the acidity or alkalinity of a solution.*)
2. What is the pH scale based on? (*USE pg. 86 - The pH scale is based on the number of hydrogen ions a solution contains. The lower the pH, the greater the concentration of hydrogen ions.*)
3. What is the pH scale? (*USE pg. 86 - The pH scale goes from 1 to 14. A pH of 7 means the solution is neutral. A pH below 7 is acidic. The closer to 0, the stronger the acid. A pH above 7 is alkaline. The closer to 14, the stronger the base.*)
4. What do indicators show? (*USE pg. 86 - Indicators are a different color at a certain pH, so they can tell the scientist whether the solution is acidic or alkaline.*)
5. What is the difference between strong acids and weak acids? (*UIDS pg. 152 - Strong acids ionize completely, while weak acids only partially ionize in solution.*) Is the same true for bases? (*UIDS pg. 152 - Yes, weak alkalis only form a few hydroxide ions.*)
6. How do you measure pH? (*UIDS pg. 152 - pH is measured by using an indicator, which changes color in the presence of an acid or a base.*)

Want More

⚗ **Soil pH** – The hydrangea plant is known for its ability to change color in various types of soil. The blooms of this plant can be blue in strongly acidic soil, purple in slightly acidic to neutral soil and pink in slightly basic soil. Have the students grow a colored hydrangea in a pot. Once it begins to bloom you can add ½ cup (150 g) of garden sulfur over the soil beneath the hydrangea before watering it to turn the blooms blue. To turn the blooms pink, do the same with powdered lime.

Sketch Week 22

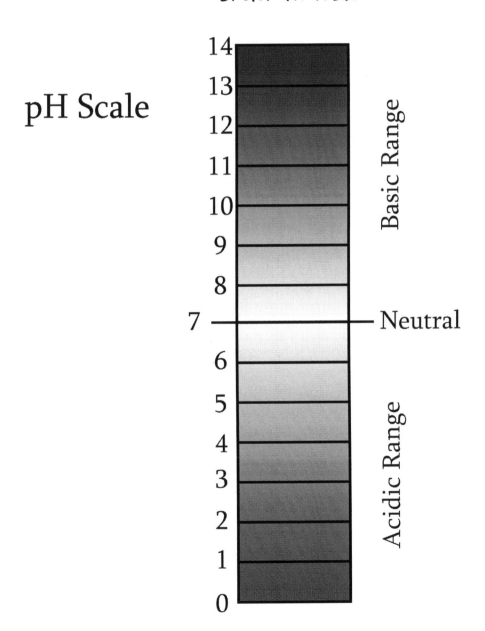

Chemistry Unit 5: Acids and Bases ~ Week 22: Measuring Acidity (pH)

Student Assignment Sheet Week 23
Neutralization and Salts

Experiment: Can I neutralize an acid?

Materials:
- ✓ Vinegar
- ✓ Ammonia
- ✓ Red cabbage solution (*Use the solution you made in week 21.*)
- ✓ Water
- ✓ Safety glasses
- ✓ 4 cups

Procedure:

1. Read the introduction to this experiment and answer the question.
2. Make the solutions using the following directions:
   - ☑ **Acidic Solution** – Mix ¼ cup (60 mL) of vinegar with ⅓ cup (79 mL) of water and label the cup "Acid".
   - ☑ **Basic Solution** – Mix ¼ cup (60 mL) of ammonia with 2 TBSP (30 mL) of water and label the cup "Base".
3. Next, mix ¼ cup (60 mL) of the acidic solution with ¼ cup (60 mL) of the red cabbage solution, which will act as an indicator and label the cup "test". In another cup, mix ¼ cup (60 mL) of water with ¼ cup (60 mL) of the red cabbage solution and label the cup "control". (**Note**—*The color of the test solution should be pink and the color of the control solution should be purple-blue.*)
4. Next, slowly add the basic solution 1 tsp (5 mL) at a time into the test solution until you achieve the color of the neutral control solution.
5. Record how many drops it took to neutralize the test solution, draw conclusions and complete the experiment sheet.

Vocabulary & Memory Work
- ☐ Vocabulary: neutralization, salts, titration
- ☐ Memory Work—This week, add the following elements to what you are working on memorizing:
  - ✓ 89-Ac-Actinium, 90-Th-Thorium, 91-Pa-Protactinium, 92-U-Uranium

Sketch: Neutralization Equation
- ☒ Label the following: acid, hydrochloric acid, base, sodium hydroxide, salt, sodium chloride, water

Writing
- ✍ Reading Assignment: *Usborne Illustrated Dictionary of Science* pp. 153-155 (Salts)
- ✍ Additional Research Readings:
  - 📖 Salts: *USE* pp. 88-89

Dates
- 🕐 1923 – Danish chemist Johannes Bronsted and English chemist Martin Lowry both suggest a change to Arrhenius' definition of acids and bases. They define an acid as a chemical that donates hydrogen ions (or protons) and a base as one that accepts them.

Schedules for Week 23
Two Days a Week

Day 1	Day 2
☐ Do the "Can I neutralize an acid?" experiment, then fill out the experiment sheet on SG pp. 166-167 ☐ Define neutralization, salts, and titration on SG pg. 145 ☐ Enter the dates onto the date sheets on SG pp. 8-13	☐ Read pp. 153-155 from *UIDS,* then discuss what was read ☐ Label the "Acid/Base Equation" sketch on SG pg. 165 ☐ Prepare an outline or narrative summary, write it on SG pp. 168-169

Supplies I Need for the Week
✓ Vinegar, Ammonia
✓ Red cabbage solution (*use the solution made in week 21*)
✓ Water, Eye dropper
✓ 4 cups

Things I Need to Prepare

Five Days a Week

Day 1	Day 2	Day 3	Day 4	Day 5
☐ Do the "Can I neutralize an acid?" experiment, then fill out the experiment sheet on SG pp. 166-167 ☐ Enter the dates onto the date sheets on SG pp. 8-13	☐ Read pp. 153-155 from *UIDS,* then discuss what was read ☐ Write an outline on SG pg. 168	☐ Define neutralization, salts, and titration on SG pg. 145 ☐ Label the "Acid/Base Equation" sketch on SG pg. 165	☐ Read one or all of the additional reading assignments ☐ Write a report from what you learned on SG pg. 169	☐ Complete one of the Want More Activities listed **OR** ☐ Study a scientist from the field of Chemistry ☐ Take the Unit 5 Test

Supplies I Need for the Week
✓ Vinegar, Ammonia
✓ Red cabbage solution (*use the solution made in week 21*)
✓ Water, Eye dropper
✓ 4 cups

Things I Need to Prepare

Additional Information Week 23

Experiment Information

☞ **Introduction** – (*from the Student Guide*) Neutralization is the process by which a solution is made neither acidic nor alkaline. Generally, a scientist will use titration to achieve the neutralization of an acid or a base. For this method, he or she will use an indicator solution that changes color with changing pH. This way the scientist will know when he or she has come close to a neutral pH of 7. In this experiment, you will use titration to attempt to reach a neutral pH by combining an acidic solution with a basic solution.

☞ **Results** – The students should see that it took around 7 to 8 tsp (35-40 mL) to neutralize the test solution. (***Note**—The white vinegar solution you made had a pH around 2.5 and the ammonia solution you made has a pH around 11.5.*)

☞ **Explanation** – Acids are chemical substances that donate hydrogen ions, or protons, to a solution and bases as ones that accept them. So, when acids and bases are combined, they tend to cancel one another out and form a salt plus water, which is known as neutralization in chemistry. In the lab, chemists use acid/base titrations to determine the concentration of an unknown acid or base. Sometimes these reactions can be violent because they can release carbon dioxide or another gas that might be toxic into the environment. So, it is a good rule to never mix an acid and base unless you are told to do so.

☞ **Take it Further** – Repeat the experiment, only this time begin with the basic solution plus the indicator. Then, slowly add the acidic solution 1 tsp (5 mL) at a time to it to achieve neutralization. Did it take the same amount? (*It should take around 8 to 9 tsp (40-45 mL) to neutralize the test solution. This is because the ammonia solution is a slightly stronger base when compared to the acidic vinegar solution.*)

Discussion Questions

1. What are salts? (*UIDS pg. 153 - A salt is an ionic compound composed of a metal cation and an anion. It is also called an acidic radical.*)
2. What is the difference between normal salts, acid salts, and base salts? (*UIDS pp. 153-154 - A normal salt is one containing a metal ion and an acidic radical. An acid salt is a salt containing hydrogen ions along with metal ions and acidic radicals. A basic salt is a salt containing a metal oxide or hydroxide along with metal ions and acidic radicals.*)
3. How are salts made? (*UIDS pg. 155 - A salt can be made through neutralization when an acid and a base are mixed. A salt can also be made through double decomposition when two or more ionic compounds react as solutions. A salt can also be made through direct synthesis and direct replacement.*)

Want More

✎ **Concentration and Neutralization** – Concentration is one of the factors that determines how much base is needed to neutralize an acid. Repeat the experiment with varying concentrations of the acidic solutions. The students can change the concentrations of this solution by adding more or less water. For example, they can use full strength vinegar for a stronger acidic solution and ¼ cup (60 mL) of vinegar plus ½ (120 mL) of water for a weaker version. (*The students should see that the stronger the concentration of the acidic solution, the more of the basic solution it will take to be neutralized.*)

Sketch Week 23

Neutralization Equation

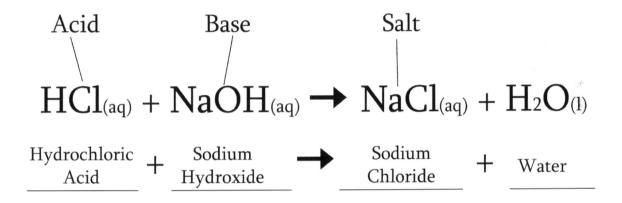

Acid   Base   Salt

$$HCl_{(aq)} + NaOH_{(aq)} \rightarrow NaCl_{(aq)} + H_2O_{(l)}$$

Hydrochloric Acid + Sodium Hydroxide → Sodium Chloride + Water

Chemistry Unit 5: Acids and Bases
Unit Test Answers

Vocabulary Matching

1. D
2. G
3. A
4. C

5. F
6. H
7. K
8. B

9. L
10. E
11. J
12. I

True or False

1. True
2. False (*Organic acids are acidic compounds produced by living things.*)
3. True
4. False (*All alkalis dissolve in water to form hydroxide ions, forming alkaline solutions.*)
5. False (*Strong acids ionize completely, while weak acids only partially ionize.*)
6. True
7. True
8. True

Short Answer

1. When acids dissolve in water they yield positively charged hydrogen ions.
2. Bases neutralize acids by accepting hydrogen ions.
3. The pH scale goes from 1 to 14. A pH of 7 means the solution is neutral. A pH below 7 is acidic. The closer to 0, the stronger the acid. A pH above 7 is alkaline. The closer to 14, the stronger the base.
4. Students need to include one of these ways: A salt can be made through neutralization when an acid and a base are mixed. A salt can also be made through double decomposition when two or more ionic compounds react as solutions. A salt can also be made through direct synthesis and direct replacement.
5. 77-Ir-Iridium, 78-Pt-Platinum, 79-Au-Gold, 80-Hg-Mercury, 81-Tl-Thallium, 82-Pb-Lead, 83-Bi-Bismuth, 84-Po-Polonium, 85-At-Astatine, 86-Rn-Radon, 87-Fr-Francium, 88-Ra-Radium, 89-Ac-Actinium, 90-Th-Thorium, 91-Pa-Protactinium, 92-U-Uranium

Chemistry Unit 5: Acids and Bases
Unit Test

Vocabulary Matching

1. Acid ___

2. Acidic Solution ___

3. Dissociation ___

4. Alkali ___

5. Alkaline Solution ___

6. Base ___

7. Indicator ___

8. Buffer ___

9. pH ___

10. Titration ___

11. Neutralization ___

12. Salt ___

A. The process by which a substance is split up into its ions in a solution.

B. A solution that resists changes in pH.

C. A base that dissolves in water to form a hydroxide ion.

D. A hydrogen containing compound that splits in water to give hydrogen ions.

E. A method of finding the concentration of acidic or alkaline solutions.

F. A solution that contains a base, or a solution that has a pH greater than seven.

G. A solution that contains an acid, or a solution that has a pH less than seven.

H. A compound that reacts with an acid to produce water and a salt.

I. A type of compound that is formed when an acid and a base react.

J. The process by which you make a solution neither acidic nor alkaline.

K. A substance that changes color as the pH of the solution changes.

L. A measure of the acidity or alkalinity of a solution.

True or False

1. _____ Acids always break down carbonates to form carbon dioxide.

2. _____ Organic acids are acidic compounds made from minerals and nonminerals.

3. _____ Bases can be used to dissolve dirt for cleaning or resin in paper making.

4. _____ All alkalis dissolve in water to form hydrogen ions, forming acidic solutions.

5. _____ Weak acids ionize completely, while strong acids only partially ionize.

6. _____ pH shows the acidity or alkalinity of a solution.

7. _____ A salt is an ionic compound composed of a metal cation and an anion.

8. _____ A normal salt is one containing a metal ion and an acidic radical.

Short Answer

1. What happens to acids in water?

2. What do bases do to acids?

3. What is the pH scale?

4. Name one way to make a salt.

5. Fill in elements 77-92 and their abbreviations from the periodic table.

> - Radon
> - Iridium
> - Astatine
> - Gold
> - Mercury
> - Protactinium
> - Lead
> - Bismuth
> - Polonium
> - Francium
> - Radium
> - Actinium
> - Platinum
> - Uranium
> - Thallium
> - Thorium

77. _____

78. _____

79. _____

80. _____

81. _____

82. _____

83. _____

84. _____

85. _____

86. _____

87. _____

88. _____

89. _____

90. _____

91. _____

92. _____

Chemistry: Unit 6

Chemistry of Life

Unit 6: Chemistry of Life
Overview of Study

Sequence of Study

Week 24: Organic Chemistry
Week 25: Enzymes
Week 26: Chemistry of Food
Week 27: Alcohols

Materials by Week

Week	Materials
24	Sugar, Salt, Candle, 2 metal spoons, Hot mitt
25	2 slices of bread, Water, Saliva, 2 plastic bags
26	Benedict's solution, Iodine solution, Several different types of food for testing (such as a hard-boiled egg, bread, potato, pasta, yogurt, cookies or cheese), Eyedropper, Small plastic cups, Safety glasses
27	Yeast, Water, Sugar, 3 bottles, 3 balloons, Instant read thermometer, Pot, Hot mitt

Vocabulary for the Unit

1. **Isomer** – A compound that contains the same number of atoms as another compound, but in a different arrangement.
2. **Monomers** – Small molecules that can join together to form long-chain polymers.
3. **Organic Compound** – A compound that contains the element carbon.
4. **Polymer** – A substance with a long-chain of molecules, formed from smaller molecules called monomers.
5. **Anabolism** – The synthesis of complex molecules from smaller ones; occurs in living things and is known as creative metabolism.
6. **Catabolism** – The breaking down of complex molecules into smaller ones; occurs in living things and is known as destructive metabolism.
7. **Enzyme** – A catalyst used by living things to increase the speed of a natural chemical process.
8. **Carbohydrates** – A group of organic compounds that consists of carbon, hydrogen, and oxygen.
9. **Lipids** – A group of solid esters that are stored in living things as a source of reserve energy, also known as fats.
10. **Minerals** – A group of inorganic compounds that are an essential components to the body's chemical processes.
11. **Protein** – A natural polymer that is made up of amino acids.

12. **Starch** – A natural polymer that is made up of glucose monomers.
13. **Vitamins** – A group of organic compounds that are essential for the normal growth and nutrition of living things.
14. **Alcohol** – A series of organic compounds in which a hydroxyl group is bound to a carbon atom; they all have the general formula $C_nH_{n+1}OH$.
15. **Fermentation** – A chemical reaction in which a sugar is broken down by microbes into alcohol and carbon dioxide.
16. **Yeast** – A single-celled fungus used in fermentation.

Memory Work for the Unit

The Elements of the Periodic Table – The following elements will be memorized in this unit:

- ✓ 93-Np-Neptunium
- ✓ 94-Pu-Plutonium
- ✓ 95-Am-Americium
- ✓ 96-Cm-Curium
- ✓ 97-Bk-Berkelium
- ✓ 98-Cf-Californium
- ✓ 99-Es-Einsteinium
- ✓ 100-Fm-Fermium
- ✓ 101-Md-Mendelevium
- ✓ 102-No-Nobelium
- ✓ 103-Lr-Lawrencium
- ✓ 104-Rf-Rutherfordium
- ✓ 105-Db-Dubnium
- ✓ 106-Sg-Seaborgium
- ✓ 107-Bh-Bohrium
- ✓ 108-Hs-Hassium

The Carbon Cycle

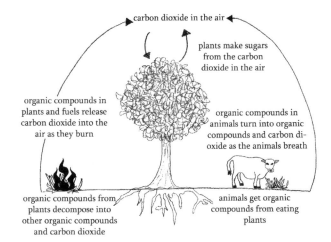

Notes

Student Assignment Sheet Week 24
Organic Chemistry

Experiment: Is there carbon in sugar or salt?

Materials:

- ✓ Sugar
- ✓ Salt
- ✓ Candle
- ✓ 2 metal spoons
- ✓ Hot mitt

> ☹ **CAUTION**
>
> **DO NOT DO THIS EXPERIMENT WITHOUT ADULT SUPERVISION!**
> This experiment involves an open flame. Be sure to use the proper safety gear.

Procedure:

1. Read the introduction to this experiment and make a hypothesis.
2. After putting on safety goggles and protective gloves, light your candle and let it burn for a few minutes.
3. Meanwhile, add about ½ tsp (2.5 g) of sugar to one of the spoons. Then, using a pair of tongs, hold the end of the spoon with the sugar over the flame for 7 minutes. Observe what happens and record what you see.
4. Then, add about ½ tsp (2.8 g) of salt to the other spoon. Using a pair of tongs, hold the end of the spoon with the salt over the flame for 7 minutes. Observe what happens and record what you see.
5. Draw conclusions and complete the experiment sheet.

Vocabulary & Memory Work

- ☐ Vocabulary: isomer, monomers, organic compounds, polymer
- ☐ Memory Work—This week, work on memorizing the carbon cycle (*see the sketch answers for details*) and add the following elements to what you are working on memorizing:
  - ✓ 93-Np-Neptunium, 94-Pu-Plutonium, 95-Am-Americium, 96-Cm-Curium

Sketch: Carbon Cycle

- 🖾 Label the following: carbon dioxide in the air, organic compounds from plants decompose into other organic compounds and carbon dioxide, plants make sugars from the carbon dioxide in the air, animals get organic compounds from eating plants, organic compounds in plants and fuels release carbon dioxide into the air as they burn, organic compounds in animals turn into organic compounds and carbon dioxide as the animals breath

Writing

- ↝ Reading Assignment: *Usborne Science Encyclopedia* pp. 92-93 (Organic Chemistry)
- ↝ Additional Research Readings:
  - 📖 Organic Chemistry: *KSE* pp. 174-175, *UIDS* pp. 190-191
  - 📖 Carbon: *KSE* pg. 170, *USE* pp. 50-53, *UIDS* pp. 178-179

Dates

- 🕐 1808 – Jons Berzelius, a Swedish chemist, first uses the term organic chemistry to refer to the chemistry of living things.
- 🕐 1828 – The meaning of organic chemistry changes to refer to the chemistry of carbon when Friedrich Wohler succeeds in synthesizing a natural carbon compound in the lab.

Schedules for Week 24
Two Days a Week

Day 1	Day 2
☐ Do the "Is there carbon in sugar or salt?" experiment, then fill out the experiment sheet on SG pp. 176-177 ☐ Define isomer, monomers, organic compounds, and polymer on SG pg. 172 ☐ Enter the dates onto the date sheets on SG pp. 8-13	☐ Read pp. 92-93 from *USE,* then discuss what was read ☐ Color and label the "Carbon Cycle" sketch on SG pg. 175 ☐ Prepare an outline or narrative summary, write it on SG pp. 178-179

Supplies I Need for the Week
- ✓ Sugar, Salt
- ✓ Candle
- ✓ 2 metal spoons
- ✓ Hot mitt

Things I Need to Prepare

Five Days a Week

Day 1	Day 2	Day 3	Day 4	Day 5
☐ Do the "Is there carbon in sugar or salt?" experiment, then fill out the experiment sheet on SG pp. 176-177 ☐ Enter the dates onto the date sheets on SG pp. 8-13	☐ Read pp. 92-93 from *USE,* then discuss what was read ☐ Write an outline on SG pg. 178	☐ Define isomer, monomers, organic compounds, and polymer on SG pg. 172 ☐ Color and label the "Carbon Cycle" sketch on SG pg. 175	☐ Read one or all of the additional reading assignments ☐ Write a report from what you learned on SG pg. 179	☐ Complete one of the Want More Activities listed **OR** ☐ Study a scientist from the field of Chemistry

Supplies I Need for the Week
- ✓ Sugar, Salt
- ✓ Candle
- ✓ 2 metal spoons
- ✓ Hot mitt

Things I Need to Prepare

Chemistry Unit 6: Chemistry of Life ~ Week 24: Organic Chemistry

Additional Information Week 24

Note

🐾 **Further Study** – Hydrocarbons, alkanes, and alkenes will be studied further in the next unit.

Experiment Information

☞ **Introduction** – (*from the Student Guide*) Organic compounds are substances that contain carbon. They generally have low melting and boiling points, which means that these compounds burn fairly easily in the air and always give off carbon dioxide. If there is incomplete combustion, which normally occurs in a non-laboratory environment, some elemental carbon or soot will remain. On the other hand, inorganic compounds typically have high melting and boiling points, which means that they require a fair amount of heat energy before they will burn. In this week's experiment, you are going to use heat to test two different common household chemicals for the presence of carbon.

☞ **Results** – The students should see that the sugar melts, gives off a gas and leaves a black residue. The students should see that the salt appears to dry out, but no other change occurs.

☞ **Explanation** – Sugar, an organic chemical with a chemical formula of $C_6H_{12}O_6$, has a melting point of 366° F (186°C). So, when the heat of the candle is added, it readily melts and as it continues to heat up, combustion occurs. In this process the sugar combines with the oxygen in the air and decomposes into carbon dioxide (gas) and water (liquid), which is shown in the reaction below:

$$O_2 + C_6H_{12}O_6 \longrightarrow CO_2 + H_2O$$

In our experiment, the combustion of the sugar was incomplete which caused a black residue or soot to be left behind. This soot is composed mainly of carbon atoms, which let us verify the presence of carbon in the sugar molecules. On the other hand, salt, an inorganic chemical with a chemical formula of NaCl, has a melting point of 1474° F (801°C). So, when the heat of the candle is added, virtually no change occurs.

☞ **Take it Further** – Try burning small samples of other kitchen compounds, such as flour, bread or cheese. You will want to do this outside, as it might start to smell badly. (*The student should see that most of the substances in the kitchen contain carbon.*)

Discussion Questions

1. What is organic chemistry? (*USE pg. 92 - Organic chemistry is the study of carbon compounds.*)
2. What is unique to organic compounds? (*USE pg. 92 - Organic compounds are compounds made from carbon, hydrogen, and oxygen that are held together by covalent bonds.*)
3. What types of bonds can carbon make? (*USE pg. 92 - Carbon can make covalent bonds. It can form single bonds where one pair of electrons is shared between two atoms, double bonds where two pairs of electrons are shared between two atoms, or triple bonds where three pairs of electrons are shared between two atoms.*)
4. What is the different between unsaturated and saturated organic compounds? (*USE pg. 93 - Unsaturated organic compounds have a double or triple bond that can open up and join with other atoms. Saturated organic compounds are full and have no free bonds that can join with other atoms.*)

5. What are synthetic compounds? (*USE pg. 93 - Synthetic compounds are ones made by chemists in a lab that copy substances that occur naturally.*)

Want More

⌂ **Building Organic Compounds** – Use marshmallows or gumdrops and toothpicks to make some organic compounds. You can make a long chain hydrocarbon, a hydrocarbon with a functional group, or cyclohexane. (***Note**—See Usborne Illustrated Dictionary of Science pp. 190-191 for pictures of these molecules.*)

⌂ **Research Report** – Have the students create a poster or chart that compares the organic and inorganic uses of carbon.

Sketch Week 24

The following sketch is not depicted in the text, but the same sketch that the students are assigned to memorize for this unit's memory work.

The Carbon Cycle

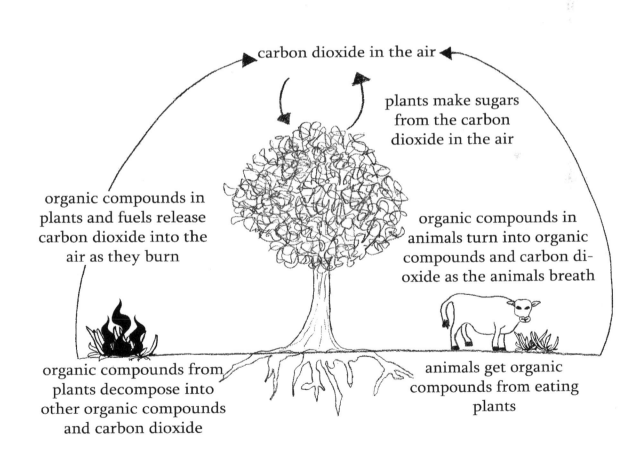

Student Assignment Sheet Week 25
Enzymes

Experiment: Does saliva digest food?

Materials:
- ✓ 2 slices of bread
- ✓ Water
- ✓ Saliva
- ✓ 2 plastic bags

Procedure:

1. Read the introduction to this experiment and make a hypothesis.
2. Label the bags #1 and #2 and then place a slice of bread into each of the bags.
3. Add ¼ cup (60 mL) of water to bag #1, seal it and set it on the counter where it won't be disturbed.
4. Spit into a cup several times, until you have about 2 TBSP (30 mL) of saliva. Add 2 TBSP (30 mL) of water and mix well. Pour the mixture over the bread in bag #2, seal the bag and set in on the counter next to bag #1.
5. Check the bags every 15 minutes over 4 hours and record what you see happening. Each time you check on the bread, be sure to quantify the amount of bread that has been broken down. For example, you could assign a percentage or use a scale, e.g. 1 (*not broken down*) to 10 (*completely broken down*).
6. After two hours, draw conclusions and complete the experiment sheet.

Vocabulary & Memory Work

- ☐ Vocabulary: anabolism, catabolism, enzyme
- ☐ Memory Work—This week, continue to work on memorizing the carbon cycle and add the following elements to what you are working on memorizing:
    - ✓ 97-Bk-Berklium, 98-Cf-Californium, 99-Es-Einsteinium, 100-Fm-Fermium

Sketch: Enzymes at Work

- ▣ Label the following: reacting molecules, enzyme, reaction occurs, product, active site

Writing

- ⌒ Reading Assignment: *Usborne Illustrated Dictionary of Science* pp. 332-333 (Metabolism)
- ⌒ Additional Research Readings:
    - 📖 Enzymes: *KSE* pg. 177
    - 📖 Energy for Life: *USE* pp. 360-361

Dates

- ⏲ 1833 – The first enzyme diastase, which we call amylase today, is discovered by Anselme Payen.
- ⏲ 1950 to today – The field of biochemistry, the study of the chemistry within living things, has greatly advanced with the development of new technology such as chromatography, X-ray diffraction, NMR spectroscopy, radio-isotopic labeling, electron microscopy, and molecular dynamics' simulations.

Schedules for Week 25
Two Days a Week

Day 1	Day 2
☐ Do the "Does saliva digest food?" experiment, then fill out the experiment sheet on SG pp. 182-183 ☐ Define anabolism, catabolism, and enzyme on SG pg. ☐ Enter the dates onto the date sheets on SG pp. 8-13	☐ Read pp. 332-333 from *UIDS,* then discuss what was read ☐ Color and label the "Enzymes at Work" sketch on SG pg. 181 ☐ Prepare an outline or narrative summary, write it on SG pp. 184-185

Supplies I Need for the Week
✓ 2 slices of bread
✓ Water
✓ Saliva
✓ 2 plastic bags

Things I Need to Prepare

Five Days a Week

Day 1	Day 2	Day 3	Day 4	Day 5
☐ Do the "Does saliva digest food?" experiment, then fill out the experiment sheet on SG pp. 182-183 ☐ Enter the dates onto the date sheets on SG pp. 8-13	☐ Read pp. 332-333 from *UIDS,* then discuss what was read ☐ Write an outline on SG pg. 184	☐ Define anabolism, catabolism, and enzyme on SG pg. 172 ☐ Color and label the "Enzymes at Work" sketch on SG pg. 181	☐ Read one or all of the additional reading assignments ☐ Write a report from what you learned on SG pg. 185	☐ Complete one of the Want More Activities listed **OR** ☐ Study a scientist from the field of Chemistry

Supplies I Need for the Week
✓ 2 slices of bread
✓ Water
✓ Saliva
✓ 2 plastic bags

Things I Need to Prepare

Additional Information Week 25

Experiment Information

☞ **Introduction** – (*from the Student Guide*) Digestion is a collection of chemical processes that occur in the body. Through these processes the food you eat is broken down into the organic compounds that your body needs. Digestion is performed within several of the organs found in your body and it uses different enzymes as catalysts to speed up the action along the way. In this week's experiment, you are going to look at one of your body's fluids to see if it aides in digesting your food.

☞ **Results** – The student should see that the bread with the saliva breaks down quicker and more completely than the bread with no saliva.

☞ **Explanation** – Saliva, which is found in the mouth, serves to moisten food and protect the teeth against certain types of bacteria. This liquid contains an enzyme, called amylase, which helps to break down starch into more easily digested sugars. This is why when you chew up bread or other starch rich foods, they taste slightly sweet. Amylase is not the only enzyme used in the digestive process, there are 7 enzymes altogether that aid in breaking down our food. In addition to salivary amylase, pancreatic amylase from the pancreas and maltase from the intestines also help to break down starches in our food. The digestion of proteins in our food is assisted by pepsin in the stomach, trypsin in the pancreas, and peptidase in the intestine. Finally, the process of breaking down the fats in our food is sped up by lipase from the pancreas.

☞ **Troubleshooting** – If the students have problems collecting enough saliva, have them keep their mouth open to prevent swallowing. Then, have them think about or smell some of their favorite food, which will increase their bodies' saliva production.

☞ **Take it Further** – Prepare another bag with a slice of bread in it. Then, add ¼ (60 mL) of cola over the bread in bag #3, seal the bag and set in on the counter. Observe what happens to the bread over 4 hours. (*The student should see that the bread was broken down even further than the bread in bags #1 and #2. This is because the cola is acidic and more closely represents the conditions in the stomach.*)

Discussion Questions

1. What are the two parts of metabolism? (*UIDS pg. 332 - The two parts of metabolism are catabolism and anabolism.*)
2. What is the difference between catabolic and anabolic reactions? (*UIDS pg. 332 - Catabolism breaks down molecules and releases energy, while anabolism use up energy to synthesized proteins, fats, and other complex substances.*)
3. What is metabolic rate, and what can cause it to change? (*UIDS pg. 332 - Metabolic rate is the overall rate at which metabolic reactions occur. Factors such as stress or temperature can slow down or speed up metabolic rate.*)
4. What are enzymes, and how does our body use them? (*UIDS pg. 333 - Enzymes are biological catalysts. Our body uses them to break down our food.*)
5. What is a kilojoule? (*UIDS pg. 333 - A kilojoule is the amount of heat energy that is produced by the catabolism of food.*)

Chemistry Unit 6: Chemistry of Life ~ Week 25: Enzymes

Want More

✎ **Research Report** – Have the students research more about the enzymes in the body. In the experiment explanation, I have already mentioned the seven digestive enzymes. However, enzymes also help to construct, synthesize, deliver and eliminate many of the chemicals that are essential for our daily functions. Have the students choose a process, such as digestion or the synthesis of DNA, and research the enzymes involved. Then, have them write a report on what they have learned. The students' final paper should include an overview of the process and what the enzymes do.

Sketch Week 25

The following sketch is not depicted in the text, but it is the students should be able to figure it out based on their reading and their knowledge of catalysts. If they struggle with this sketch, use the answers below to explain it to them.

Enzymes at Work

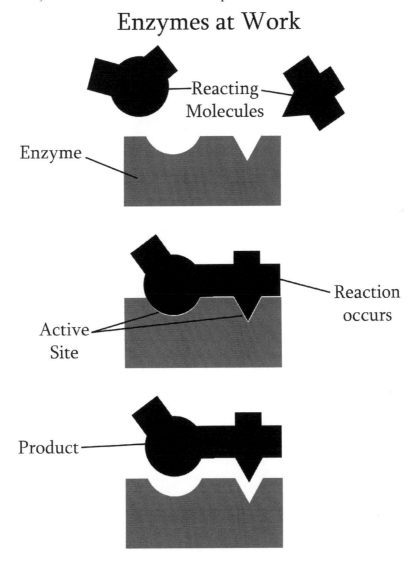

Student Assignment Sheet Week 26
Chemistry of Food

Experiment: Is it a sugar or a starch?

Materials:

- ✓ Benedict's solution, Iodine swab
- ✓ Several different types of food for testing (such as a hard-boiled egg, bread, potato, pasta, yogurt, cookies or cheese)
- ✓ Eyedropper
- ✓ Small plastic cups
- ✓ Safety glasses

> ☹ **CAUTION:**
> The chemicals used in this experiment will stain clothing. Please use protective gear or wear old clothing when performing this experiment.

Procedure:

1. Read the introduction to this experiment and write down which of the test foods you think contain starch and which ones you think contain sugar.
2. Test each of your sample foods using the two tests below to see if they contain starch or sugar. (**Note**—*Directions for making the sample solution for the sugar test are on the experiment sheet on SG pg. 188.*)
   - ☑ **Starch test** – Cut off a sample of the food, then use an iodine swab to rub some solution onto it. If the iodine changes color to blue, then starch is present. If it remains the same brown color, no starch is present. Write down your results in the chart on your experiment sheet. Then, repeat this procedure for each of the test samples.
   - ☑ **Sugar Test** – Add 40 drops of the sample solution to a small plastic cup and put on your safety glasses. Add 10 drops of Benedict's solution and mix well. If the solution turns green, yellow or brick-red, sugar is present in some concentration. If the solution remains blue, there is no sugar in the sample solution.
3. Draw conclusions and complete the experiment sheet.

Vocabulary & Memory Work

- ☐ Vocabulary: carbohydrates, lipids, minerals, protein, starch, vitamins
- ☐ Memory Work—This week, add the following elements to what you are working on memorizing:
   - ✓ 101-Md-Mendelevium, 102-No-Nobelium, 103-Lr-Lawrencium, 104-Rf-Rutherfordium

Sketch: Glucose Breakdown

- ▣ Label the following: glucose, oxygen, carbon dioxide, water, energy

Writing

- ✎ Reading Assignment: *Usborne Illustrated Dictionary of Science* pp. 204-205 (Food)
- ✎ Additional Research Readings:
   - 📖 Food and Nutrition: *KSE* pg. 126
   - 📖 Food and Diet: *USE* pp. 356-357

Dates

- 🕐 1813 – The first book on food chemistry, entitled *Elements of Agricultural Chemistry*, is published by Sir Humphrey Davy.

Schedules for Week 26
Two Days a Week

Day 1	Day 2
☐ Do the "Is it a sugar or a starch?" experiment, then fill out the experiment sheet on SG pp. 188-189 ☐ Define carbohydrates, lipids, minerals, protein, starch, and vitamins on SG pg. 172-173 ☐ Enter the dates onto the date sheets on SG pp. 8-13	☐ Read pp. 204-205 from *UIDS,* then discuss what was read ☐ Color and label the "Glucose Breakdown" sketch on SG pg. 187 ☐ Prepare an outline or narrative summary, write it on SG pp. 190-191

Supplies I Need for the Week
✓ Benedict's solution, Iodine swab
✓ Several different types of food for testing (such as a hard-boiled egg, bread, potato, pasta, yogurt, cookies or cheese)
✓ Eyedropper, Small plastic cups, Safety glasses

Things I Need to Prepare

Five Days a Week

Day 1	Day 2	Day 3	Day 4	Day 5
☐ Do the "Is it a sugar or a starch?" experiment, then fill out the experiment sheet on SG pp. 188-189 ☐ Enter the dates onto the date sheets on SG pp. 8-13	☐ Read pp. 204-205 from *UIDS,* then discuss what was read ☐ Write an outline on SG pg. 190	☐ Define carbohydrates, lipids, minerals, protein, starch, and vitamins on SG pg. 172-173 ☐ Color and label the "Glucose Breakdown" sketch on SG pg. 187	☐ Read one or all of the additional reading assignments ☐ Write a report from what you learned on SG pg. 191	☐ Complete one of the Want More Activities listed **OR** ☐ Study a scientist from the field of Chemistry

Supplies I Need for the Week
✓ Benedict's solution, Iodine swab
✓ Several different types of food for testing (such as a hard-boiled egg, bread, potato, pasta, yogurt, cookies or cheese)
✓ Eyedropper, Small plastic cups, Safety glasses

Things I Need to Prepare

Additional Information Week 26

Experiment Information

☞ **Introduction** – (*from the Student Guide*) There is a whole branch of chemistry dedicated to the science of food. Scientists in this field study the chemical processes that are associated with food production and preservation. Some of the most basic tests that these chemists use are to determine whether or not starch or sugar is present in a given sample of food. In this week's experiment, you are going to perform these tests on several food samples from your own kitchen.

☞ **Results** – The students' answers will vary depending upon the substances that they chose to test.

☞ **Explanation** – This experiment was designed to give your students a firsthand look at how scientists test unknown foods in a laboratory setting. If they are able to gather results, they will have achieved the goals of this experiment.

☞ **Take it Further** – Test the same sample foods for the presence of protein and fat. You will need Biuret solution and Sudan III stain.

☑ **Protein test** – Cut off a sample of the food and mash it up well. Add ½ cup (120 mL) of water to it and mix well. Add 40 drops of the sample solution to a small plastic cup and put on your safety glasses. Add 3 drops of Biuret solution and mix well. If the solutions turns pink or purple, protein is present. If the solution remains blue, there is no protein in the sample solution. Write down your results in the chart on your experiment sheet. Then, repeat this procedure for each of your test samples.

☑ **Fat (Lipid) Test** – Add 20 drops of the sample solution to a small plastic cup and 20 drops of water. With your safety glasses on, add 3 drops of Sudan III stain and mix well. If a red-stained oil layer separates out, there is fat present in the solution. If not, there is no fat in the sample solution.

Discussion Questions

1. What are the groups of chemicals that are essential to life? (*UIDS pg. 204 - The groups of chemicals that are essential to life are proteins, carbohydrates, fats, vitamins, minerals, and water. We also need fiber to help move food through the gut.*)

2. What is the difference between a monosaccharide and a polysaccharide? (*UIDS pg. 204 - A monosaccharide, such as glucose, is formed from one organic compound unit. A polysaccharide, such as starch, is formed from many individual organic compound units.*)

3. What are amino acids, and what do they make? (*UIDS pg. 205 - Amino acids are molecules that contain a carbon atom joined to a carboxyl group and an amino group. Amino acids are the building blocks of proteins.*)

4. What do animals need proteins for, and where can they get it? (*UIDS pg. 205 - Animals need proteins for repairing and growing tissue. Protein is found in meat, dairy food, nuts, cereal, and beans.*)

5. What do animals need vitamins for? (*UIDS pg. 205 - Animals need vitamins to help enzyme reactions in the body.*)

Want More

⌂ **Food Preservation** – Have the students test potential food preservatives using apples. Cut several apples in half, leave one untreated and then treat the rest with substances that the students believe will prevent browning. (***Note***—*Let the students lead in choosing the substances, but be sure to guide them to lemon juice as one of the options as it will preserve the apple. This will give you a good comparison for the unpreserved control.*)

Sketch Week 26

Glucose Breakdown

$$C_6H_{12}O_6 + 6O_2 \rightarrow 6CO_2 + 6H_2O + 2900 \text{ kJ}$$

glucose   oxygen   carbon dioxide   water   energy

Student Assignment Sheet Week 27
Alcohols

Experiment: Does temperature affect fermentation?

Materials:

- ✓ Yeast
- ✓ Water
- ✓ Sugar
- ✓ 3 bottles
- ✓ 3 balloons
- ✓ Instant read thermometer
- ✓ Pot
- ✓ Hot mitt

> ☣ **CAUTION**
> **DO NOT DO THIS EXPERIMENT WITHOUT ADULT SUPERVISION!**
> Boiling hot water can cause severe damage. Be sure to use the proper safety gear.

Procedure:

1. Read the introduction to this experiment and make a hypothesis.
2. Add 1 TBSP (8.5 g) of yeast and 1 TBSP (15 g) of sugar into each of the bottles. Label them #1 through #3.
3. Fill a pot with 4 cups of water and attach the instant read thermometer to the side. Slowly heat the water. Once it reaches 75° F (24° C), use the hot mitt to pour out one cup (240 mL) of water and then replace the pot on the burner. Quickly add the cup of 75° F (24° C) water to bottle #1, swirl the contents until mixed and place the balloon over the top.
4. Once the water reaches 95° F (35° C), use the hot mitt to pour out one cup (240 mL) of water and then replace the pot on the burner. Quickly add the cup of 95° F (35° C) water to bottle #2, swirl the contents until mixed and place the balloon over the top.
5. Once the water reaches 135° F (57° C), use the hot mitt to pour out one cup (240 mL) of water and then replace the pot on the burner. Quickly add the cup of 135° F (57° C) water to bottle #3, swirl the contents until mixed and place the balloon over the top.
6. After 15 minutes, measure the circumference of each of the balloons and record the measurements.
7. Draw conclusions and complete the experiment sheet.

Vocabulary & Memory Work

- ☐ Vocabulary: alcohol, fermentation, yeast
- ☐ Memory Work—This week, add the following elements to what you are working on memorizing:
  - ✓ 105-Db-Dubnium, 106-Sg-Seaborgium, 107-Bh-Bohrium, 108-Hs-Hassium

Sketch: Fermentation

🖼 Label the following: stopper keeps out oxygen; carbon dioxide gas forms; glucose is broken down and ethanol is made; yeast, glucose, and water is mixed; glucose, enzyme added, ethanol, carbon dioxide

Writing

☜ Reading Assignment: *Usborne Illustrated Dictionary of Science* pp. 196-197 (Alcohols)

Dates

- 🕐 3000 BC – Yeast is used to make alcoholic drinks, such as beer or wine through fermentation.
- 🕐 200 – Yogurt is made through the use of bacteria.

Chemistry Unit 6: Chemistry of Life ~ Week 27: Alcohols

Schedules for Week 27
Two Days a Week

Day 1	Day 2
☐ Do the "Does temperature affect fermentation?" experiment, then fill out the experiment sheet on SG pp. 194-195 ☐ Define alcohol, fermentation, and yeast on SG pg. 173 ☐ Enter the dates onto the date sheets on SG pp. 8-13	☐ Read pp. 196-197 from *UIDS,* then discuss what was read ☐ Color and label the "Fermentation" sketch on SG pg. 193 ☐ Prepare an outline or narrative summary, write it on SG pp. 196-197

Supplies I Need for the Week
- ✓ Yeast, Water, Sugar
- ✓ 3 bottles, 3 balloons
- ✓ Instant read thermometer
- ✓ Pot, Hot mitt

Things I Need to Prepare

Five Days a Week

Day 1	Day 2	Day 3	Day 4	Day 5
☐ Do the "Does temperature affect fermentation?" experiment, then fill out the experiment sheet on SG pp. 194-195 ☐ Enter the dates onto the date sheets on SG pp. 8-13	☐ Read pp. 196-197 from *UIDS,* then discuss what was read ☐ Write an outline on SG pg. 196	☐ Define alcohol, fermentation, and yeast on SG pg. 173 ☐ Color and label the "Fermentation" sketch on SG pg. 193	☐ Read one or all of the additional reading assignments ☐ Write a report from what you learned on SG pg. 197	☐ Complete one of the Want More Activities listed **OR** ☐ Take the Unit 6 Test

Supplies I Need for the Week
- ✓ Yeast, Water, Sugar
- ✓ 3 bottles, 3 balloons
- ✓ Instant read thermometer
- ✓ Pot, Hot mitt

Things I Need to Prepare

Additional Information Week 27

Experiment Information

☞ **Introduction** – (*from the Student Guide*) Fermentation is the process in which the sugar in fruits or grains is converted by microbes into alcohol and carbon dioxide. Some of the most common fermentations are done by a group of microbes known as yeasts. Yeasts act as catalysts in the conversion of the glucose present in the fruit or grain to alcohol and carbon dioxide, which is shown in the equation below.

$$C_6H_{12}O_6 \text{ (glucose)} \longrightarrow 2C_2H_5OH \text{ (ethanol)} + 2CO_2 \text{ (carbon dioxide)}$$

These fermentation processes require the right conditions to maximize their yield. In this week's experiment, you will test to see if temperature affects the amount of carbon dioxide the yeast can produce.

☞ **Results** – The students should see that the balloon on top of bottle #2 inflated the most, followed by balloon on bottle #1 and then the balloon on bottle #3.

☞ **Explanation** – Yeast speeds up the fermentation process; in other words, it acts as a catalyst to the conversion of glucose into ethanol and carbon dioxide. This microbial catalyst has an optimum temperature range of 85° F (29° C) to 100° F (38° C). The ideal temperature for yeast is 95° F (35° C), which is why bottle #2 produced the most carbon dioxide. If the temperature is below this optimum range, the yeast will not be fully activated. Since bottle #1 was at 75° F (24° C), the yeast did not fully activate, which is why it produced a low amount of carbon dioxide. If the temperature is above this optimum range, the yeast will begin to die. Since bottle #3 was at 135° F (57° C), the yeast was dying off, which is why it produced a low amount of carbon dioxide. If the temperature was allowed to reach 140° F (60° C), all the yeast in the solution would die.

☞ **Take it Further** – Try the experiment again, but this time allow the student to choose two different temperatures, one below 95° F (35° C) and one above 95° F (35° C). (*The students' results will vary based on the temperatures that they choose.*)

Discussion Questions

1. What are alcohols? (*UIDS pg. 196 - Alcohols are organic compounds with one or more hydroxyl groups.*)
2. Why do short-chain alcohols mix with water, but long-chain alcohols do not? (*UIDS pg. 196 - Short-chain alcohols mix with water because they are slightly polar. Long-chain alcohols do not mix with water as they are less polar.*)
3. Why is yeast used in alcohol fermentation? (*UIDS pg. 197 - Yeast in used in alcohol fermentation because it has the enzyme zymase, which catalyzes the reaction that changes glucose to ethanol.*)
4. What is a condensation reaction? (*UIDS pg. 197 - A condensation reaction is one in which two molecules react together to form one molecule with the loss of water.*)

Want More

✐ **Make your own Yogurt** – A common fermentation process is the one used to make yogurt. One of the easiest ways to make homemade yogurt is in the crock pot. You will need ½ gallon

(1.9 L) of milk and ½ cup (122 g) of plain yogurt (*plain Greek yogurt gives the best results*). Pour the milk into your crock pot and cook it on low for 2 ½ hours. Unplug the crock pot off and let it sit undisturbed for 3 hours. Then pull out 1 cup (120 mL) of the milk and mix it with ½ cup (122 g) of the plain yogurt. Pour this mixture back into the crock pot and replace the lid. Cover the crock pot with a thick towel and let it sit undisturbed overnight (*or 8 to 12 hours*). The next morning you will have your own smooth and delicious homemade yogurt! You can strain this to make it thicker, if desired. Here's the chemistry behind the yogurt making process:

🍴 *Milk contains lactose, which is a sugar. This sugar is fermented by a bacteria called Lactobacillus, which is in the starter. The fermentation process produces lactic acid, which causes the proteins in the milk to coagulate. The end result is a mixture with thick consistency and slightly sour taste, which we call yogurt.*

Sketch Week 27

Fermentation

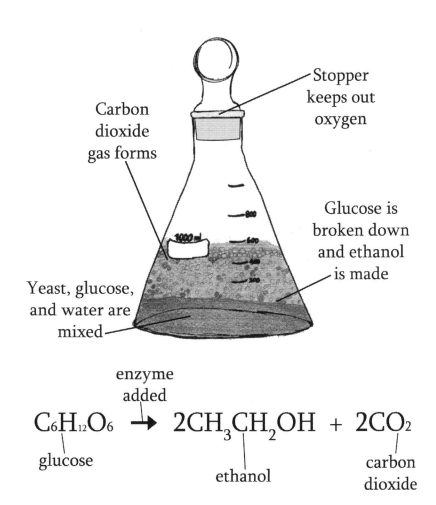

Stopper keeps out oxygen

Carbon dioxide gas forms

Glucose is broken down and ethanol is made

Yeast, glucose, and water are mixed

enzyme added

$$C_6H_{12}O_6 \rightarrow 2CH_3CH_2OH + 2CO_2$$

glucose   ethanol   carbon dioxide

Unit 6: Chemistry of Life
Unit Test Answers

Vocabulary Matching

1. F
2. B
3. L
4. C
5. E
6. A
7. P
8. H
9. O
10. I
11. K
12. M
13. G
14. J
15. N
16. D

True or False

1. False (*Unsaturated organic compounds have a double or triple bond that can open up and join with other atoms.*)
2. True
3. True
4. False (*Catabolic reactions release energy, while anabolic reactions use up energy.*)
5. True
6. False (*A polysaccharide, such as starch, is formed from many individual organic compound units.*)
7. True
8. False (*Short-chain alcohols mix with water because they are slightly polar. Long-chain alcohols do not mix with water.*)

Short Answer

1. The students should have at least five of the following groups of chemicals that are essential to life: proteins, carbohydrates, fats, fiber, vitamins, minerals, and water.
2. Enzymes are biological catalysts. Our body uses them to break down our food.
3. See the carbon cycle sketch from week 24.
4. Yeast in used in alcohol fermentation because it has the enzyme zymase, which catalyzes the reaction that changes glucose to ethanol.
5. 93-Np-Neptunium, 94-Pu-Plutonium, 95-Am-Americium, 96-Cm-Curium, 97-Bk-Berkelium, 98-Cf-Californium, 99-Es-Einsteinium, 100-Fm-Fermium, 101-Md-Mendelevium, 102-No-Nobelium, 103-Lr-Lawrencium, 104-Rf-Rutherfordium, 105-Db-Dubnium, 106-Sg-Seaborgium, 107-Bh-Bohrium, 108-Hs-Hassium

Unit 6: Chemistry of Life
Unit Test

Vocabulary Matching:

1. Isomer ___

2. Monomers ___

3. Organic Compound ___

4. Polymer ___

5. Anabolism ___

6. Catabolism ___

7. Enzyme ___

8. Carbohydrates ___

9. Lipids ___

10. Minerals ___

11. Protein ___

12. Starch ___

13. Vitamins ___

14. Alcohol ___

15. Fermentation ___

16. Yeast ___

A. The breaking down of complex molecules into smaller ones; occurs in living things and is known as destructive metabolism.

B. Small molecules that can join together to form long-chain polymers.

C. A substance with a long-chain of molecules, formed from smaller molecules called monomers.

D. A single-celled fungus used in fermentation.

E. The synthesis of complex molecules from smaller ones; occurs in living things and is known as creative metabolism.

F. A compounds that contain the same number of atoms as another compound, but in a different arrangement.

G. A group of organic compounds that are essential for the normal growth and nutrition of living things.

H. A group of organic compounds that consists of carbon, hydrogen, and oxygen.

I. A group of inorganic compounds that are an essential component to the body's chemical processes.

J. A series of organic compounds in which a hydroxyl group is bound to a carbon atom; they all have the general formula $C_nH_{n+1}OH$.

K. A natural polymer that is made up of amino acids.

L. A compound that contains the element carbon.

M. A natural polymer that is made up of glucose monomers.

N. A chemical reaction in which a sugar is broken down by microbes into alcohol and carbon dioxide.

O. A group of solid esters that are stored in living things as a source of reserve energy, also known as fats.

P. A catalyst used by living things to increase the speed of a natural chemical process.

True or False

1. _____ Saturated organic compounds have a double or triple bond that can open up and join with other atoms.

2. _____ Organic compounds are compounds make from carbon, hydrogen, and oxygen that are held together by covalent bonds.

3. _____ A kilojoule is the amount of heat energy that is produced by the catabolism of food.

4. _____ Anabolic reactions release energy, while catabolic reactions use up energy.

5. _____ Animals need proteins for repairing and growing tissue.

6. _____ A monosaccharide, such as glucose, is formed from many individual organic compound units.

7. _____ A condensation reaction is one in which two molecules react together to form one molecule with the loss of water.

8. _____ Long-chain alcohols mix with water because they are slightly polar. Short-chain alcohols do not mix with water.

Short Answer

1. Name at least five of the groups of chemicals that are essential to life.

2. What are enzymes, and how does our body use them?

3. Label the following sketch of the carbon cycle.

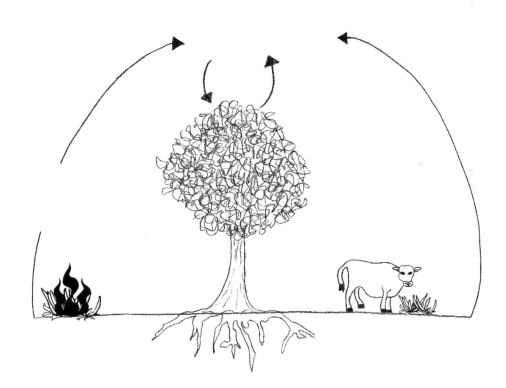

184

4. Why is yeast used in alcohol fermentation?

5. Fill in elements 93-108 from the periodic table.

- Seaborgium
- Curium
- Neptunium
- Mendelevium
- Americium
- Einsteinium
- Dubnium
- Berkelium
- Lawrencium
- Californium
- Bohrium
- Fermium
- Plutonium
- Hassium
- Rutherfordium
- Nobelium

93. _____

94. _____

95. _____

96. _____

97. _____

98. _____

99. _____

100. _____

101. _____

102. _____

103. _____

104. _____

105. _____

106. _____

107. _____

108. _____

Chemistry: Unit 7

Chemistry of Industry

Unit 7: Chemistry of Industry
Overview of Study

Sequence of Study

Week 28: Soaps and Detergents
Week 29: Alkanes and Alkenes
Week 30: Homologous Groups
Week 31: Petrochemicals
Week 32: Polymers and Plastics
Week 33: Iron and Alloys
Week 34: Radioactivity
Week 35: Pollution

Note—*This unit also contains a science fair project and a scientist biography project for the students to complete.*

Materials by Week

Week	Materials
28	Powdered detergent, Liquid soap, 2 large cups, 2 small cups, 2 bowls, pH paper, Vegetable oil, Dirt, Ketchup, Plaster of Paris, Water, Straw, Old T-shirt fabric
29-32	Materials will vary depending on the Science Fair Project that your student has chosen to do
33-35	*No experiment supplies needed.*

Vocabulary for the Unit

1. **Detergent** – A synthetic or organic substance, not prepared from a fat, that helps to remove grease and oil in water.
2. **Soap** – A metallic salt derived from a fat that is used to help break up grease and oil so that it can be removed.
3. **Alkane** – A class of saturated hydrocarbons.
4. **Alkene** – A class of unsaturated hydrocarbons with at least one carbon-to-carbon double bond.
5. **Ester** – A homologous series of organic compounds that give fruits and flowers their fragrances.
6. **Homologous Series** – A group of organic compounds that have the same general formula for all its members.
7. **Hydrocarbon** – A chemical compound that consists of only hydrogen and carbon.
8. **Natural Polymer** – A polymer that is produced by a living organism, such as cellulose, starch, chitin, and proteins.

9. **Synthetic Polymer** – A polymer that is man-made, also known as plastics.
10. **Alloy** – A mixture of two or more metals, or a metal, and a nonmetal.
11. **Raw Materials** – The basic materials used to make a product.
12. **Slag** – Impurities found in iron ore that are left over from the smelting process.
13. **Radioactive Decay** – The process by which a nucleus ejects particles by radiation until stability is reached.
14. **Radioisotope** – An unstable nucleus that has a different number of neutrons than a stable nucleus.
15. **Biodegradable** – A term used to describe matter that can be broken down into simpler substances by bacteria.
16. **Pollutant** – A substance that was released into the atmosphere, oceans, or rivers that upset the natural processes of the Earth.

Memory Work for the Unit

The Elements of the Periodic Table – The following elements will be memorized in this unit:
- ✓ 109-Mt-Meitnerium
- ✓ 110-Ds-Darmstadium
- ✓ 111-Rg-Roentgenium
- ✓ 112-Cn-Copernicium

Note—Students can also memorize elements 113-118. These are extremely unstable and very little is known about these elements. So, for the purpose of this study, I have omitted them from the required memory work list. Here are the current (as of 2016) names:
- ✓ 113-Nh-Nihonium
- ✓ 114-Fl-Flerovium
- ✓ 115-Mc-Moscovium
- ✓ 116-Lv-Livermorium
- ✓ 117-Ts-Tennessine
- ✓ 118-Og-Oganesson

Notes

Student Assignment Sheet Week 28
Soaps and Detergents

Experiment: Will soap and detergent always act the same?
Materials:

- ✓ Powdered detergent
- ✓ Liquid soap
- ✓ 2 large cups
- ✓ 2 small cups
- ✓ 2 bowls
- ✓ pH paper
- ✓ Vegetable oil
- ✓ Dirt
- ✓ Ketchup
- ✓ Plaster of Paris
- ✓ Water
- ✓ Straw
- ✓ Old T-shirt fabric

Procedure:

1. Read the introduction to this experiment and make a hypothesis.
2. Begin by making your solutions. First, mix 1 cup (240 mL) of water with 1 ½ tsp (5 g) of powdered detergent in a cup. Label the cup with "detergent solution". Then, mix 1 cup (240 mL) of water with 1 ½ tsp (7.5 mL) of liquid soap. Label the second cup with "soap solution".
3. Pour ½ cup (120 mL) of each solution into two different shallow bowls. Then, use the pH paper to test and record the pH of each of your solutions. Next, rub three pieces of T-shirt fabric with the dirt and ketchup. Set one of the pieces aside and place the other two in each of the bowls. Set aside to soak for 30 minutes.
4. Now add ¼ cup (60 mL) of each of the solutions to the small cups. Then, add 1 tsp (5 mL) of vegetable oil to each and agitate the solution. Observe and record what happened. Pour the mixtures out and rinse the cups well.
5. Add ¼ cup (60 mL) of each of the solutions to the small cups. Then, add 1 TBSP (10 g) of Plaster of Paris to each and agitate the solution. Using a straw blow into each of the solutions for 10 seconds, observe and record what happened. Let the solutions sit for 30 minutes and then observe what has happened.
6. After 30 minutes, remove the T-shirt fabric from each of the bowls, agitate by rubbing for 1-2 minutes and rinse well. Compare the two washed pieces with the unwashed one, observe and record your results.
7. Draw conclusions and complete the experiment sheet.

Vocabulary & Memory Work
- ☐ Vocabulary: detergent, soap
- ☐ Memory Work—Work on memorizing the following elements:
    - ✓ 109-Mt-Meitnerium, 110-Ds-Darmstadium, 111-Rg-Roentgenium, 112-Cn-Copernicium

Sketch: Micelle
- ▦ Label the following: grease, detergent molecules, hydrophilic end of detergent molecule sticks out of grease, hydrophobic end of detergent molecule embeds in grease

Writing
- ⤶ Reading Assignment: *Usborne Illustrated Dictionary of Science* pp. 202-203 (Detergents)
- ⤶ Additional Research Readings:
    - 📖 Soaps and Detergents: *KSE* pg. 187, *USE* pg. 94 (section on detergents)

Chemistry Unit 7: Chemistry of Industry ~ Week 28: Soaps and Detergents

Schedules for Week 28
Two Days a Week

Day 1	Day 2
☐ Do the "Will soap and detergent always act the same?" experiment, then fill out the experiment sheet on SG pp. 204-205 ☐ Define detergent and soap on SG pg. 200 ☐ Enter the dates onto the date sheets on SG pp. 8-13	☐ Read pp. 202-203 from *UIDS*, then discuss what was read ☐ Color and label the "Micelle" sketch on SG pg. 203 ☐ Prepare an outline or narrative summary, write it on SG pp. 206-207

Supplies I Need for the Week
- ✓ Powdered detergent, Liquid soap
- ✓ 2 large cups, 2 small cups, 2 bowls
- ✓ pH paper, Vegetable oil, Dirt, Ketchup, Plaster of Paris
- ✓ Water, Straw, Old T-shirt fabric

Things I Need to Prepare

Five Days a Week

Day 1	Day 2	Day 3	Day 4	Day 5
☐ Do the "Will soap and detergent always act the same?" experiment, then fill out the experiment sheet on SG pp. 204-205 ☐ Enter the dates onto the date sheets on SG pp. 8-13	☐ Read pp. 202-203 from *UIDS*, then discuss what was read ☐ Write an outline on SG pg. 206	☐ Define detergent and soap on SG pg. 200 ☐ Color and label the "Micelle" sketch on SG pg. 203	☐ Read one or all of the additional reading assignments ☐ Write a report from what you learned on SG pg. 207	☐ Complete one of the Want More Activities listed **OR** ☐ Study a scientist from the field of Chemistry

Supplies I Need for the Week
- ✓ Powdered detergent, Liquid soap
- ✓ 2 large cups, 2 small cups, 2 bowls
- ✓ pH paper, Vegetable oil, Dirt, Ketchup, Plaster of Paris
- ✓ Water, Straw, Old T-shirt fabric

Things I Need to Prepare

Additional Information Week 28

Notes

🖈 **Memory Work**—Students can also memorize elements 113-118. These are extremely unstable and very little is known about these elements. So, for the purpose of this study, I have omitted them from the memory work list.

Experiment Information

☞ **Introduction** – (*from the Student Guide*) Detergents and soaps are a class of molecules that assist water in removing grease and oil from dirty material. Detergents are synthetic or organic substances that are not prepared from fats, while soaps are metallic salts derived from fats. These molecules have a water-loving, polar head and a grease-loving, non-polar tail. So, they are able to dissolve in water, but at the same time they can attract grease and remove it from a surface. In this experiment, you are going to perform several tests to see how detergents and soaps work.

☞ **Results** – The students should see the following:

	pH	Mixing with Oil	Hard Water Test (Plaster of Paris)	Cleanliness Test
Soap Solution	Will vary with brand	Forms an emulsion	Forms some bubbles, after 30 minutes the bubbles disappear and scum forms	Removes some of the stain
Detergent Solution	Will vary with brand	Forms an emulsion	Forms a lot of bubbles, after 30 minutes the bubbles have started to disappear	Removes almost all of the stain

☞ **Explanation** – Soaps and detergents are two different types of chemicals that both assist in removing dirt and oil from materials. Both have a water-loving head and a grease-loving tails, but the do differ in their functionality. Soaps react with the minerals in the hard water to form scum, which keeps them from forming too many bubbles, while detergents do not. Soaps are also not nearly as effective in cleaning cloth because they need quite a bit more water to do so, while detergents need less water to clean cloth and are far mor gently on these materials. The differences between these two classes of chemicals are a direct result of their chemical make-up.

☞ **Take it Further** – Try repeating the experiment with hot water and in ice cold water. (*You should see that soap performs better in hot water, while detergent performs the same in hot and cold water.*)

Discussion Questions

1. What are the three ways detergents work? (*UIDS pg. 202 - Detergents work by lowering water's surface tension, by enabling grease molecules to dissolve in water, and by keeping*

the removed dirt suspended in the water.)

2. How do soaps and detergents differ? (*UIDS pg. 202 - Soaps are a type of detergent prepared from fats. Soaps can create scum, while soapless detergents typically do not.*)
3. Describe a detergent molecule. (*UIDS pg. 202 - Detergent molecules have a hydrophilic (water-loving) head with a functional group that makes the end polar. They also have a hydrophobic (grease-loving) tail made of a long chain hydrocarbon.*)
4. What are synthetic detergents? (*UIDS pg. 203 - Synthetic detergents are soapless detergents made from the by-products of refining crude oil.*)
5. What makes a detergent biodegradable? (*UIDS pg. 203 - A detergent is biodegradable if it can be broken down by bacteria in nature.*)
6. What do surfactants do? (*UIDS pg. 203 - Surfactants lower the surface tension of water.*)

Want More

⌂ **Make your own soap** – The easiest way to make soap is by using the cold-process method. You will need oil (*such as vegetable or olive oil*), Lye (*100% sodium hydroxide*), water, thermometer, several bowls and molds. The following video is a great tutorial on soap making because it includes an explanation of the chemistry involved.

🖥 http://www.youtube.com/watch?v=1MtzyxQiqKo

Sketch Week 28

Micelle

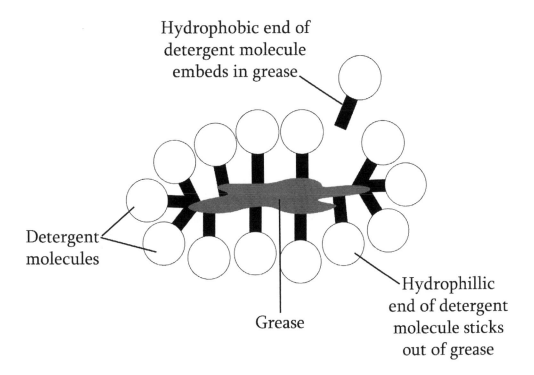

Hydrophobic end of detergent molecule embeds in grease.

Detergent molecules

Grease

Hydrophillic end of detergent molecule sticks out of grease

Student Assignment Sheet Week 29
Alkanes and Alkenes

Science Fair Project

This week, you will complete step one and begin step two of your Science Fair Project. You will be choosing your topic, formulating a question and doing some research about that topic.

1. **Choose your topic** – You should choose a topic in the field of chemistry that interests you, such as electrolysis. Next, come up with several questions you have relating to that topic, (e.g. "Does the salt concentration affects electrolysis?" or "Does the strength of the battery affect the speed of electrolysis?"). Then, choose the one question you would like to answer and refine it (e.g. "Does salt affect the speed of hydrogen production in electrolysis?").

2. **Do Some Research** – Now that you have a topic and a question for your project, it is time to learn more about your topic so that you can make an educated guess (hypothesis) on the answer to your question. For the question stated above, you would need to research electrolysis and how hydrogen gas is produced during the process. Begin by looking up the topic in the references you have at home. Then, make a trip to the library to search for more on the topic. As you do your research, write any relevant facts you have learned on index cards and be sure to record the sources you use.

Vocabulary & Memory Work

- [ ] Vocabulary: alkane, alkene
- [ ] Memory Work—Review the elements of the periodic table that you have memorized this year.

Sketch: Hydrogenation

- Label the Following: ethene, ethane, unsaturated compound, saturated compound, carbon-to-carbon single bond, carbon-to-carbon double bond

Writing

- Reading Assignment: *Usborne Illustrated Dictionary of Science* pg. 192 (Alkanes), pg. 193 (Alkenes)
- Additional Research Readings:
  - Alkanes and Alkenes: *USE* pp. 96-96

Dates to Enter

- No dates to be entered.

Schedules for Week 29
Two Days a Week

Day 1	Day 2
☐ Define alkane and alkene on SG pg. 200 ☐ Read pp. 192-193 from *UIDS*, then discuss what was read ☐ Color and label the "Hydrogenation" sketch on SG pg. 209 ☐ Prepare an outline or narrative summary, write it on SG pp. 212-213	☐ Decide on a topic and question for your Science Fair Project, record them on SG pg. 210 ☐ Research your Science Fair Project topic, see SG pg. 210-211 for details ☐ Enter the dates onto the date sheets on SG pp. 8-13

Supplies I Need for the Week
✓ Index Cards

Things I Need to Prepare

Five Days a Week

Day 1	Day 2	Day 3	Day 4	Day 5
☐ Read pp. 192-193 from *UIDS*, then discuss what was read ☐ Prepare an outline or narrative summary, write it on SG pp. 212	☐ Define alkane and alkene on SG pg. 200 ☐ Color and label the "Hydrogenation" sketch on SG pg. 209	☐ Read one or all of the additional reading assignments ☐ Prepare your report, write the report on SG pg. 213 ☐ Enter the dates onto the date sheets on SG pp. 8-13	☐ Decide on a topic and question for your Science Fair Project, record them on SG pg. 210 ☐ Begin the research for your Science Fair Project topic, see SG pg. 210-211 for details	☐ Begin the research for your Science Fair Project topic, see SG pg. 210-211 for more information

Supplies I Need for the Week
✓ Index Cards

Things I Need to Prepare

Additional Information Week 29

Science Fair Project

- **Step 1: Choose the Topic** — The students will be choosing a topic for their science fair project this week. Have them choose a topic in the field of chemistry that interests them. You can use Janice VanCleave's *A+ Science Fair Projects* and Janice VanCleave's *A+ Projects in Chemistry: Winning Experiments for Science Fairs and Extra Credit*.

  1. **Key 1 ~ Decide on an area of chemistry.** The students should choose an area that fascinates them, something in chemistry that they want to know more about. You will begin by leading the students to brainstorm about things in chemistry that interest them. Have them rank these areas by degree of interest and then choose one area on which to focus. If their area is too broad, you will want them to narrow it down a bit. You can do this by asking them what they find interesting about the particular field.

  2. **Key 2 ~ Develop several questions about the area of chemistry.** Once the students have determined their area of chemistry, they need to develop several questions about their topic that they can answer with their project. Good questions begin with how, what, when, who, which, why or where. At this point, you are just getting them to think of possible questions.

  3. **Key 3 ~ Choose a question to be the topic.** Now that the students have several options of questions that they can answer with their science fair project, you will need to have them choose one of those questions for their project. Some of their questions will be easy to develop into an experiment for their science fair project that will determine the answer' some will not. If you think that their question is too broad, you need to help them narrow it down to something more specific.

- **Step 2: Do Some Research** — The students will also begin researching their topics. You may need to walk them through this process if they have not had much experience with doing research prior to this.

  1. **Key 1 ~ Brainstorm for research categories.** This is an important key, because developing relevant research categories before they begin to search for information will help the students to maintain a more focused approach. It will also help the students know where to begin their research and how to determine what information is important to their project and what is not. Keep in mind that some students may have a harder time coming up with categories that relate to their topic, so you may need to give them additional assistance. The students should have at least three categories and no more than five. This will help them to obtain relevant information as well as make it easier for them to write their report. Once the students have chosen their research categories, have them assign each category a number.

  2. **Key 2 ~ Research the categories.** Depending upon their experience with research, you may or may not have to walk the students through this entire process. Either way, have them begin by looking at the reference material that they have close at hand, such as encyclopedias that they own or that are in the classroom. Then, they can look to their local library or the Internet for additional information. As the students uncover bits of relevant data, have them write each fact in their own words on a separate index card. They should number each card at the top left with the category in which it fits, which will make them easier to organize. We also recommend that they assign a letter for each reference they use, which they can write in the right-hand top corner of each card. This way, after they organize and sort their cards,

they will know which references they need to include in their bibliography. See the following article for more information on the index card system:

🖳 http://elementalblogging.com/the-index-card-system/

Discussion Questions

1. What are some of the common properties of alkanes? (*UIDS pg. 192 - Alkanes are non-polar molecules that burn in air to form carbon dioxide and water. They also react with halogens, but beyond that they are considered unreactive.*)

2. What happens to the state of alkanes and alkenes as the molecules get longer? (*UIDS pp. 192-193 - As alkanes and alkenes get longer, they change state from gas to liquid.*)

3. What is a substitution reaction? (*UIDS pg. 192 - A substitution reaction is one where an atom or functional group is replaced by a different atom or functional group*)

4. What are some of the common properties of alkenes? (*UIDS pg. 193 - Alkenes burn with a smoky flame. In an excess of oxygen, alkenes oxidize to carbon dioxide and water. They are more reactive than alkanes because of the double bond.*)

5. What is an addition reaction? (*UIDS pg. 193 - An addition reaction is one where two molecules react to produce a single larger molecule.*)

Want More

↻ **Hydrogenation Comparison** – Have the students observe and compare vegetable oil and margarine. (*The oil is an unsaturated compound that undergoes hydrogenation to become the saturated compound known as margarine.*)

Sketch Week 29

Hydrogenation

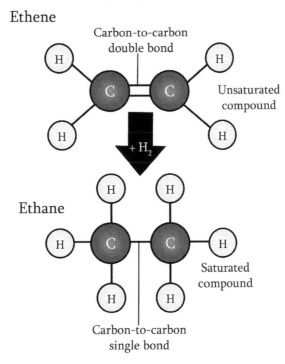

Chemistry Unit 7: Chemistry of Industry ~ Week 29: Alkanes and Alkenes

Student Assignment Sheet Week 30
Homologous Groups

Science Fair Project

This week, you will complete steps two through four of your Science Fair Project. You will be finishing your research, formulating your hypothesis and designing your experiment.

2. **Do Some Research** – This week, you will finish your research. Then, organize your research index cards and write a brief report on what you have found out.

3. **Formulate a Hypothesis** – A hypothesis is an educated guess. For this step, you need to review your research and make an educated guess about the answer to your question. A hypothesis for the question asked in step one would be, "If the concentration of the salt is increased, then the rate that hydrogen gas is produced during electrolysis with increase."

4. **Design an Experiment** – Your experiment will test the answer to your question. It needs to have a control and several test groups. Your control will have nothing changed, while your test groups will change only one factor at a time. An experiment to test the hypothesis given above would be to make three different solutions for electrolysis with varying salt concentration for your test groups and use plain distilled water for the control group. Then, allow the electrolysis to run for each group until you collect a set amount of hydrogen gas. Each time, you would record the amount of time it took. If time allows, you can go ahead and begin your experiment this week.

Vocabulary & Memory Work
- ☐ Vocabulary: ester, homologous series
- ☐ Memory Work—Review the elements of the periodic table that you have memorized this year.

Sketch:
- ▨ There is no sketch this week.

Writing
- ᘒ Reading Assignment: *Usborne Illustrated Dictionary of Science* pp. 194-195 (Alkynes and More Homologous Series)
- ᘒ Additional Research Readings:
  - 📖 Esters: *USE* pg. 95

Dates to Enter:
- 🕐 No dates to be entered.

Schedules for Week 30
Two Days a Week

Day 1	Day 2
☐ Finish your research, formulate a hypothesis for your Science Fair Project and record on SG pp. 215-216 ☐ Design your experiment for your Science Fair Project, see SG pg. 216-217 for details	☐ Read pp. 194-195 from *UIDS,* then discuss what was read ☐ Prepare an outline or narrative summary, write it on SG pp. 218-219 ☐ Define ester and homologous series on SG pg. 200

Supplies I Need for the Week

Things I Need to Prepare

Five Days a Week

Day 1	Day 2	Day 3	Day 4	Day 5
☐ Finish your research and formulate a hypothesis for your Science Fair Project and record on SG pp. 215-216	☐ Design your experiment for your Science Fair Project, see SG pg. 216-217 for details	☐ Read pp. 194-195 from *UIDS,* then discuss what was read ☐ Prepare an outline or narrative summary, write it on SG pp. 218	☐ Define ester and homologous series on SG pg. 200	☐ Read one or all of the additional reading assignments ☐ Write a report from what you learned on SG pg. 219

Supplies I Need for the Week

Things I Need to Prepare

Additional Information Week 30

Notes
❦ **Sketch and Dates**—There is no sketch or dates to be entered this week in order to give more time for the students to work on their science fair project.

Science Fair Project
♀ **Step 2: Do Some Research** – This week, the students will finish their research and write a brief report for their project board.

1. **Key 3 ~ Organize the information.** Once the students have finished their research, you need to have them organize and sort through the information that they have found. Begin by having the students sort their cards into piles using the research categories which are in the top left hand of their index card. Then, have the student read through each fact and determine five to seven of the most relevant pieces of information from each pile. You will need to help them as they decide which facts are relevant to their project (i.e., useful for answering their topical question) and which ones are not. These facts will then form the basis of their report.

2. **Key 4 ~ Write a brief report.** Have the students determine the order they want to share their research categories. Normally, they would go from broad information about their subject to more specific information for their project. After they do this, they need to take the five to seven facts from the first category and turn them into a three to four sentence paragraph by combining the facts into a coherent passage. They will repeat this process until they have a three to five paragraph paper. Then, the students will need to edit and revise their paper so that it becomes a cohesive report. Finally, they will need to add in a bibliography with the resources they used for their report.

♀ **Step 3: Formulate a Hypothesis** – The students need to review their research and apply what they have learned to help them determine the answer to their question.

1. **Key 1 ~ Review the research.** The students will need to review their research, so that it is fresh in their minds. You can do this by having them read over their report, or by having them read over each of the index cards they made. The level of involvement for this key will depend on how much time goes by between step two and step three of the science fair project.

2. **Key 2 ~ Formulate an answer.** After the students have reviewed their research, they should also read over their question one more time. Once all the information is fresh in their minds, they are ready to make an educated guess at the answer to their question. Guide them to craft a response in the form of an if-then statement that they will be able to design an experiment to test. However, keep in mind that not all questions can be answered easily with if-then statements. As they design their experiment in the next step, you can still have them make a few adjustments to their hypotheses, if necessary.

♀ **Step 4: Design an Experiment** – The students will also design their experiment this week. You may need to walk them through this process by suggesting ways they could test for the answer to their question.

1. **Key 1 ~ Choose a test.** You can ask the students what kind of a test could you use that would answer your question and prove your hypothesis either true or false. Have them write down each idea they have, but keep in mind that the students may need a fair amount of help

with this process. If they find that they cannot come up with any options for testing their hypotheses, they may need to tweak their statements a bit. If they decide to do this, make sure they verify that the new versions still answer their original topical questions. Once the students have written down several ideas, have them review the options and choose one of the ideas for their projects.

2. **Key 2 ~ Determine the variables.** Now that the students have chosen a method to test their hypotheses, they need to determine the variables that will exist in their test. Here's a link to an article to help you understand the different types of variables:

   💻 http://elementalblogging.com/experiment-variables/

   Have them answer the questions found in the student guide to determine their independent, dependent and controlled variables.

3. **Key 3 ~ Plan the experiment.** Now that the students understand the variables that are at work, they are ready to use this information along with their testing idea to create an experiment design. You need to explain to them that they must have a control group as well as several test groups. The control group will have nothing changed, while each of the test groups will have only one change to the independent variable. The students should also plan on having several samples in each of their test groups. Once you have explained to the students the parameters of their experiment, they can begin formulating a plan by determining what their test groups will be. Then, they need to decide how long they have to run their tests. Once they have this information, they will write out their experiment design.

Discussion Questions

1. What are alkynes? (*UIDS pg. 195 - Alkynes are organic compounds with a triple carbon-carbon bond. They are produced by cracking and can be used to make plastics and solvents.*)
2. What is the difference between the different homologous series? (*UIDS pg. 195 - The different homologous series have different functional groups, which give them different chemical properties.*)
3. What is the functional group in ketones, and how are they used? (*UIDS pg. 195 - Ketones have a -CO- functional group. These organic compounds are mostly colorless liquids and are used in solvents.*)
4. What is the functional group in carboxylic acids, and how are they used? (*UIDS pg. 195 - Carboxylic acids have a -COOH functional group. These organic compounds are colorless weak acids that react with alcohols to form esters.*)
5. What is the functional group in esters, and how are they used? (*UIDS pg. 195 - Esters have a -COO- functional group. These organic compounds play a role in the flavors and smells of fruits and flowers. They are also found in vegetable oils and animal fats.*)

Want More

🕯 **Smell Test** – You will need a bottle of perfume. Esters, a homologous series with a -COO- functional group, play a role in the smells of fruits and flowers. Chemists synthesize these natural molecules to create perfumes and flavorings. Have the students test the smell of a perfume to try to determine which fruit or flower scents have been included.

Student Assignment Sheet Week 31
Petrochemicals

Science Fair Project

This week, you will complete steps five and six of your Science Fair Project. You will carry out the experiment and record your observations and results.

5. **Perform the Experiment** – This week, you will perform the experiment you designed last week. Be sure to take pictures along the way as well as record your observations and results. (*Note—Observations are a record of the things you see happening in your experiment. For instance, an observation would be that the hydrogen gas from the control group solution filled almost half of the test tube. Results are specific and measurable. For instance, results could be that you collected 10 cm of hydrogen gas in 20 minutes from the electrolysis test #1. Observations are generally recorded in journal form, while results can be compiled into tables, charts and graphs or relayed in paragraph form.*)

6. **Analyze the Data** – Once you have compiled your observations and results, you can use them to answer your question. You need to look for trends in your data and make conclusions from that. A possible conclusion to the electrolysis experiment would be, "The more salt present in the electrolyte solution, the quicker the hydrogen gas is produced in electrolysis." If your hypothesis does not match your conclusion or your were not able to answer your question using the results from your experiment, you may need to go back and do some additional experimentation.

Vocabulary & Memory Work

- ☐ Vocabulary: hydrocarbon
- ☐ Memory Work—Review the elements of the periodic table that you have memorized this year.

Sketch: Fractional Distillation of Crude Oil

- ▨ Label the Following: crude oil is heated, furnace, fractionating column, refinery gases, gasoline, kerosene, diesel fuel, residue

Writing

- ᕍ Reading Assignment: *Usborne Illustrated Dictionary of Science* pp. 198-199 (Petroleum)
- ᕍ Additional Research Readings:
  - 📖 Petrochemicals: *KSE* pp. 190-191
  - 📖 Crude Oil: *USE* pp. 98-99

Dates to Enter:

- 🕐 347 – Oil wells are first drilled in China. The oil was naturally refined and used as a building adhesive, for water-proofing and as a fuel.
- 🕐 1850's – The first commercial oil refinery appears.

Schedules for Week 31
Two Days a Week

Day 1	Day 2
☐ Perform the experiment for your Science Fair Project and analyze your observations and results on SG pp. 222-223 ☐ Define hydrocarbon on SG pg. 200 ☐ Enter the dates onto the date sheets on SG pp. 8-13	☐ Read pp. 198-199 from *UIDS*, then discuss what was read ☐ Color and label the "Fractional Distillation of Crude Oil" sketch on SG pg. 221 ☐ Prepare an outline or narrative summary, write it on SG pp. 224-225

Supplies I Need for the Week

Things I Need to Prepare

Five Days a Week

Day 1	Day 2	Day 3	Day 4	Day 5
☐ Perform the experiment for your Science Fair Project and record your observations and results on SG pg. 222 ☐ Enter the dates onto the date sheets on SG pp. 8-13	☐ Read pp. 198-199 from *UIDS*, then discuss what was read ☐ Write an outline on SG pg. 224	☐ Define hydrocarbon on SG pg. 200 ☐ Color and label the "Fractional Distillation of Crude Oil" sketch on SG pg. 221	☐ Read one or all of the additional reading assignments ☐ Write a report from what you learned on SG pg. 225	☐ Finish the experiment for your Science Fair Project ☐ Analyze your observations and results on SG pg. 222-223

Supplies I Need for the Week

Things I Need to Prepare

Chemistry Unit 7: Chemistry of Industry ~ Week 31: Petrochemicals

Additional Information Week 31

Science Fair Project

⚲ **Step 5: Perform the Experiment** – The students will be performing the experiment and recording their observations and results. Be sure to check in with them to see how they are doing.

1. **Key 1 ~ Get ready for the experiment.** The students already have a plan in place, but there are still a few things they need to do before beginning their experiment. They need to look at a calendar and make sure that they will be home for the duration of the trial because they will need to be there to make observations and record results on each day of testing. The students also need to gather and prep any materials that they will be using during their experiment.

2. **Key 2 ~ Run the experiment.** The students have done a lot of work to reach this point, but that preparation has paved a smooth road for their experiment. At this point, they are familiar with their research and their design, so they should be able to carry out their testing with little to no help. You want to make sure that the students write down a list of things they need to check each day during the experiment. Be sure that they include taking pictures of what they see on their list as they will need these images for their project board.

3. **Key 3 ~ Record any observations and results.** As the students run their experiment, they need to compile their observations and results. Observations are the record of the things the scientist sees happening in an experiment, while results are specific and measurable. Observations are generally recorded in journal form, while results can be compiled into tables, charts or graphs. You will need to help the students create a table to record their results as well as provide them with a journal for their observations. Once they finish their experiment, you may need to help them chart or graph their data.

⚲ **Step 6: Analyze the Data** – The students will now analyze their observations and results to draw conclusions from their experiment.

1. **Key 1 ~ Review and organize the data.** The students need to analyze their observations and results to determine if their hypotheses are true or false. To do this, you need to have them read over each of their journal entries and note any trends in their observations. You also need to have the students interpret the charts or graphs they created in the last step and write down the information that they can glean from them.

2. **Key 2 ~ State the answer.** Now that the students have noted trends from their observations and interpreted information from their results, they can use this data to answer their question. They need to first determine if they have proved their hypotheses true or false. Once the students have decided if their hypotheses statements were true or false, they can craft a one sentence answer to their original topical questions from step one. Their statements should begin with, "I found that ___" or "I discovered that ___." In the rare case that the students are unable to state an answer to their question, they need to take what they have learned, go back to the drawing table and redesign their experiment.

3. **Key 3 ~ Draw several conclusions.** When the students draw conclusions, they are putting into words what they have learned from their project. Their conclusion should include the following information:

☑ The answer to their question;

☑ Whether or not their hypotheses were proven true (***Note**—If their hypotheses were*

proven false, they should state why.);
☑ Any problems or difficulties they ran into while performing their experiment;
☑ Anything interesting they discovered that they would like to share;
☑ Ways that they would like to expand their experiment in the future.
It should be one paragraph, or about four to six sentences in length. Have the students begin their concluding paragraph with the statement they wrote for the previous key.

Discussion Questions

1. What is the difference between petroleum and natural gas? (*UIDS pg. 198 - Petroleum and natural gas are often found together, but petroleum is a mixture of alkanes that can be refined, while natural gas consists mainly of methane.*)

2. What are the three main processes in refining? Describe each. (*UIDS pg. 198 - The three main processes in refining are primary distillation, cracking, and reforming. In primary distillation, crude oil is heated and separated in a fractionating column. In cracking, the larger alkanes are broken into smaller alkanes and alkenes. In reforming, gasoline is produced by breaking up straight-chain alkanes and reassembling them as branched-chain molecules.*)

3. Describe the five main products of primary distillation. (*UIDS pg. 199 - The five main products of primary distillation are refinery gas, gasoline, kerosene, diesel oil, and residue. Refinery gas is mainly methane, along with propane and butane. Gasoline consists of alkanes with 5 to 12 carbon atoms per molecule. Kerosene consists of alkanes with 9 to 15 carbon atoms per molecule. Diesel oil consist of alkanes with 12 to 25 carbon atoms per molecule. The residue that is left can be use as lubricating oil, waxes, and asphalt.*)

Sketch Week 30

Distillation of Crude Oil

⇨ Refinery gases
⇨ Gasoline
⇨ Kerosene
⇨ Diesel fuel
⇨ Residue

Fractionating column

Crude oil is heated

Furnace

Want More

☞ **Learn More** – Watch the following video on fractional distillation:
💻 https://youtu.be/CjmriZq5xRo watch?v=CjmriZq5xRo

Chemistry Unit 7: Chemistry of Industry ~ Week 31: Petrochemicals

Student Assignment Sheet Week 32
Polymers and Plastics

Science Fair Project

This week, you will complete steps seven and eight of your Science Fair Project. You will be writing and preparing a presentation of your Science Fair Project.

7. **Create a Board** – This week, you will be creating a visual representation of your science fair project that will serve as the centerpiece of your presentation. You will begin by planning the look of your board, then move onto preparing the information and finally you will pull it all together.

8. **Give a Presentation** – After you have completed your presentation board, determine if you would like to include part of your experiment in your presentation. Then, prepare a 5 minute talk about your project, be sure to include the question you tried to answer, your hypothesis, a brief explanation of your experiment and the results plus the conclusion to your project. Be sure to arrive on time for your presentation. Set up your project board and any other additional materials. Give your talk and then ask if there are any questions. Answer the questions and end your time by thanking whoever has come to listen to your presentation.

Vocabulary & Memory Work

- ☐ Vocabulary: natural polymer, synthetic polymer
- ☐ Memory Work—Review the elements of the periodic table that you have memorized this year.

Sketch: Making a Polymer by Condensation

- Label the Following: the two monomers approach each other, a reaction occurs and several hydrogen and oxygen atoms split off, the two monomer chains join to form a polymer, the released hydrogen and oxygen atoms combine to form water

Writing

- Reading Assignment: *Usborne Illustrated Dictionary of Science* pp. 200-201 (Polymers and Plastics)
- Additional Research Readings:
    - 📖 Polymers: *KSE* pg. 205
    - 📖 Plastics: *KSE* pg. 206
    - 📖 Polymers and Plastics: *USE* pp. 100-101

Dates to Enter

- 🕐 1850's – Alexander Parkes, an English chemist, synthesizes the first polymer.
- 🕐 1965 – Stephanie Kwolek, an American chemist for DuPont, creates the first Kevlar fibers, a polymer that is used to protect law enforcement officers from bullets.

Schedules for Week 32
Two Days a Week

Day 1	Day 2
☐ Prepare your science fair project board and present your Science Fair Project, see SG pp. 228-229 for details ☐ Enter the dates onto the date sheets on SG pp. 8-13	☐ Read pp. 200-201 from *UIDS*, then discuss what was read ☐ Color and label the "Making a Polymer by Condensation" sketch on SG pg. 227 ☐ Define natural polymer and synthetic polymer on SG pp. 200-201 ☐ Prepare an outline or narrative summary, write it on SG pp. 230-231

Supplies I Need for the Week

Things I Need to Prepare

Five Days a Week

Day 1	Day 2	Day 3	Day 4	Day 5
☐ Begin working on your science fair project board, see SG pp. 228-229 for details	☐ Read pp. 200-201 from *UIDS*, then discuss what was read ☐ Write an outline on SG pg. 230-231	☐ Define natural polymer and synthetic polymer on SG pp. 200-201 ☐ Color and label the "Making a Polymer by Condensation" sketch on SG pg. 227	☐ Enter the dates onto the date sheets on SG pp. 8-13 ☐ Continue working on your science fair project board	☐ Present your Science Fair Project, see SG pg. 229 for details

Supplies I Need for the Week

Things I Need to Prepare

Chemistry Unit 7: Chemistry of Industry ~ Week 32: Polymers and Plastics

Additional Information Week 32

Science Fair Project

₽ **Step 7: Create a Board** — In this step, the students will pull together all the information they have learned to create a presentation board.

1. **Key 1 ~ Plan out the board.** The science fair project board is the visual representation of the students' hard work, so you definitely want them to put as much effort into this step as they have into the others. The board will have specific sections that are set, but they should personalize the look with color and graphics that suit their tastes and match their projects. Please see the Appendix pg. 256 for a more detailed explanation of the science fair project board layout.

2. **Key 2 ~ Prepare the information.** The students have put in a lot of effort until this point, but the work they have done in the previous steps will make it easier for them to prepare the information for their board. The students need to type the information up and choose a font and font size for their board. Please see the Appendix pg. 256 for a more detailed explanation of what each section should include.

3. **Key 3 ~ Put the board together.** Now that the students have planned out their science fair project board and prepared the information, they are ready to pull it all together. They need to cut out the decorative elements and glue them to the backboard. Then, they need to print and cut out their informational paragraphs. For added depth, they can glue the paragraphs onto a foam board before adding the information. Finally, the students should add their title and the finishing touches to their board.

₽ **Step 8: Give a Presentation** — This step gives the students a chance to communicate with an audience what they have learned from their project. The best way to achieve this is to have the students participate in a Science Fair where their projects will be judged, but if that's not possible, don't skip this key. The students can still present their project to their family or to a group of their peers.

1. **Key 1 ~ Prepare the presentation.** Once the students have finished their project board, they can begin to work on their presentation. They should prepare a brief five minute talk about their science fair project. This talk should include the question they tried to answer, their hypotheses, a brief explanation of their experiment, the results, and the conclusion to their project. You will need to guide the students as they turn their information paragraphs into an outline for their presentation. This outline should highlight the main points that they want to cover for their presentation.

2. **Key 2 ~ Practice the presentation.** Once the students have finished preparing the outline for their talk, have them practice in front of a mirror. They should practice looking at the audience while pointing to the different sections on their project board as they present. Once they feel confident with their presentation, have them give a practice talk to you. Be sure to give them feedback, so that they can make the necessary changes before they present their science fair project to a group.

3. **Key 3 ~ Share the presentation.** It is important to have the students present their work to an audience and answer related questions from the group. This will reinforce what they have learned as well as help them to discern how to communicate what they know.

Discussion Questions

1. What are polymers? (*UIDS pg. 200 - Polymers are substances made from repeating sequences of monomers bonded together.*)
2. What is the difference between a homopolymer and a copolymer? (*UIDS pg. 200 - A homopolymer is made from identical monomers, while a copolymer is made from two or more different monomers.*)
3. What are natural polymers? Give examples. (*UIDS pg. 201 - Nature polymers occur in nature, such as starch, proteins, and rubber.*)
4. What are synthetic polymers? Give examples. (*UIDS pg. 201 - Synthetic polymers are man-made, such as plastics, polyesters, polystyrene, acrylic, and PVC.*)
5. How can plastics be bad for the environment? (*UIDS pg. 201 - Plastics can be bad for the environment because they are often not biodegradable and give off toxic fumes when burned.*)

Want More

✐ **Make your own Polymer** – You will need water, borax and white glue. Dissolve 1 tsp (8.5 g) of borax in 1 cup (240 mL) of water in a bowl. In a separate cup mix ½ cup (120 mL) of white glue with ½ cup (120 mL) of water. You can add food coloring to the mixture at this point if you desire. Next, pour the glue mixture into the borax solution and gently stir. The mixture should begin to polymerize immediately, but continue to stir the mixture for as long as possible. Once the solid polymer has formed, remove it from the excess liquid and gently kneed it by hand until it is soft and pliable.

Sketch Week 32

The following sketch is not directly depicted in the text, but the information is shared. Your students may need additional help to label sketch.

Making a Polymer by Condensation

the two monomers approach each other

a reaction occurs and several hydrogen and oxygen atoms split off

the two monomer chains join to form a polymer

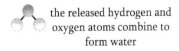

the released hydrogen and oxygen atoms combine to form water

Chemistry Unit 7: Chemistry of Industry ~ Week 32: Polymers and Plastics

Student Assignment Sheet Week 33
Iron and Alloys

Scientist Biography Report Project
This week, you will complete step one and begin step two of your Scientist Biography Report Project. You will be choosing the scientist you would like to learn more about and begin your research. The instructions for this week's assignments are on the following Scientist Biography Report sheets.

Vocabulary & Memory Work
- [ ] Vocabulary: alloy, raw materials, slag
- [ ] Memory Work—Review the elements of the periodic table that you have memorized this year.

Sketch: Blast Furnace Reaction
- Label the Following: limestone, coke, iron ore, heat, molten iron, The burning coke forms carbon monoxide which reacts with iron ore, limestone traps the impurities left in the iron ore to form slag

Writing
- Reading Assignment: *Usborne Science Encyclopedia* pp. 34-35 (Alloys) and pp. 36-37 (Iron and Steel
- Additional Research Readings:
  - Iron, Copper, and Zinc: *UIDS* pg. 174
  - Iron: *KSE* pg. 198
  - Alloys: *KSE* pp. 202-203

Dates to Enter
- c. 3000 BC – The first alloy is created during the Bronze Age.
- 1856 – Henry Bessemer invents a way of removing most of the carbon from iron, creating steel.

Schedules for Week 33
Two Days a Week

Day 1	Day 2
☐ Choose your scientist and begin your research for the Scientist Biography Report, see SG pp. 234-235 for details ☐ Enter the dates onto the date sheets on SG pp. 8-13	☐ Read pp. 34-37 from *USE*, then discuss what was read ☐ Color and label the "Blast Furnace Reaction" sketch on SG pg. 233 ☐ Define alloy, raw materials, and slag on SG pg. 201 ☐ Prepare an outline or narrative summary, write it on SG pp. 236-237

Supplies I Need for the Week

Things I Need to Prepare

Five Days a Week

Day 1	Day 2	Day 3	Day 4	Day 5
☐ Choose your scientist and begin your research for the Scientist Biography Report, see SG pp. 234-235 for details	☐ Read pp. 34-37 from *USE*, then discuss what was read ☐ Write an outline on SG pg. 236-237	☐ Define alloy, raw materials, and slag on SG pg. 201 ☐ Color and label the "Blast Furnace Reaction" sketch on SG pg. 233	☐ Enter the dates onto the date sheets on SG pp. 8-13 ☐ Continue working on your Scientist Biography Report	☐ Continue working on your Scientist Biography Report

Supplies I Need for the Week

Things I Need to Prepare

Chemistry Unit 7: Chemistry of Industry ~ Week 33: Iron and Alloys

Additional Information Week 33

Scientist Biography Report Step 1: Choose a Scientist

☞ **From the Student Guide** – During the next three weeks, you are going to be researching and learning more about a scientist that has contributed to the field of chemistry. This week, you need to begin your scientist biography project by choosing which scientist you will research. You can choose one of the scientists mentioned in the "Dates" sections or you can choose one that has interested you.

☞ **Step 1 Notes** – The students will need some help with step one, especially if they are not familiar with some of the well-known scientist in the field of chemistry. They can choose one of the more famous chemists, like Marie Curie, or one of the lesser known chemists, such as Frederick Wohler. Either way, they need to choose a person that has enough information written about him or her for the students to compile a two to three page paper. Here are several lists to help as you guide the students through choosing their topic:
  - 🖥 http://chemistry.about.com/od/historyofchemistry/a/famouschemists.htm
  - 🖥 http://library.thinkquest.org/C006439/scientists/

Scientist Biography Report Step 2: Research the Scientist

☞ **From the Student Guide** – Once you have chosen the scientist you would like to study, you can begin your research. Begin by looking for a biography on your chosen scientist at the library. Then, look for articles on the chemist in magazines, newspapers, encyclopedias or on the Internet. You will need to know the following about your scientist to write your report:
  - ☑ Biographical information on the scientist (*i.e., where they were born, their parents, siblings, and how they grew up*);
  - ☑ The scientist's education (*i.e., where they went to school, what kind of student they were, what they studied, and so on*);
  - ☑ Their scientific contributions (*i.e., research that they participated in, any significant discoveries they made, and the state of the world at the time of their contributions*).

  As you read over the material you have gathered, be sure to write down any facts you glean in your own words. You can do this on the sheet below or on separate index cards. You will have two weeks to complete your research, so plan accordingly.

☞ **Step 2 Notes** – I highly recommend having the students use the index card system to record their research findings. You can read more about this method by clicking below:
  - 🖥 http://elementalblogging.com/the-index-card-system/

Discussion Questions

1. Why are alloys better than a pure metal on its own? (*USE pg. 34 - Alloys contain a combination of metals, adding their properties and making the resulting material stronger or lighter than the pure metals alone.*)

2. What are some examples of alloys? (*USE pg. 35 - Steel is an alloy that is a mixture of iron and carbon, giving the material strength and flexibility. Brass is an alloy of copper and zinc, yielding a material that is easily shaped and resistant to corrosion. Duralumin is an alloy of aluminum and magnesium, giving the resulting material strength and resistance to corrosion.*)

3. Where is iron found, and what are the two most common iron ores? (*USE pg. 36 - Iron is found in the Earth's crust. The two most common iron ores are hematite and magnetite.*)
4. What happens when iron ore is smelted in a blast furnace? (*USE pg. 36 - Iron ore is heated up with limestone and a carbon compound known as coke. The coke burns, producing carbon monoxide and heating up the iron ore. Iron oxide in the ore reacts with the carbon monoxide, freeing the iron and forming carbon dioxide. Limestone combines with the remaining impurities to form slag.*)
5. What is steel, and why is it beneficial? (*USE pg. 37 - Steel is iron with more carbon removed and several other elements added. It is stronger than pure iron.*)

Want More
☞ **Research Report** – Have the students write a brief report on the Bronze Age and its contribution to the field of chemistry. Their paper should include information on alloys and how they were created.

Sketch Week 33
The following sketch is not directly depicted in the text, but the information is shared. Your students may need additional help to label sketch.

Blast Furnace Reaction

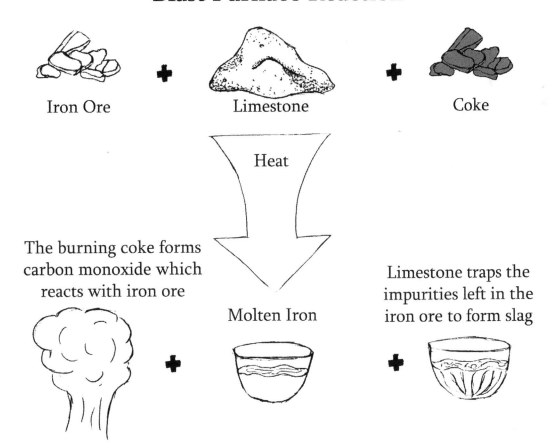

Iron Ore + Limestone + Coke

Heat

The burning coke forms carbon monoxide which reacts with iron ore

Molten Iron

Limestone traps the impurities left in the iron ore to form slag

Chemistry Unit 7: Chemistry of Industry ~ Week 33: Iron and Alloys

Student Assignment Sheet Week 34
Radioactivity

Scientist Biography Report Project

This week, you will complete step two and three of the Scientist Biography Report Project. You will be finishing up your research and organizing the information you have gathered into an outline. The instructions for this week's assignments are on the following Scientist Biography Report sheets.

Vocabulary & Memory Work

- [ ] Vocabulary: radioactive decay, radioisotope
- [ ] Memory Work—Review the elements of the periodic table that you have memorized this year.

Sketch

- There is no sketch this week.

Writing

- Reading Assignment: *Usborne Illustrated Dictionary of Science* pp. 128-129 (Radioactivity)
- Additional Research Readings:
  - 📖 Radiation: *KSE* pp. 244-245
  - 📖 Radioactivity: *USE* pp. 114-115

Dates

- 🕒 1896 – Antoine Becquerel is working with a natural fluorescent material and x-rays, which leads to his discovery of radioactivity.
- 🕒 1898 – Marie and Pierre Curie coin the term "radioactive" when they discovered radium and polonium.
- 🕒 1928 – Hans Geiger and one of his students develop a machine that can detect and measure the intensity of radiation.

Schedules for Week 34
Two Days a Week

Day 1	Day 2
☐ Finish the research and write an outline for the Scientist Biography Report, see SG pp. 239-242 for details ☐ Enter the dates onto the date sheets on SG pp. 8-13	☐ Read pp. 128-129 from *UIDS*, then discuss what was read ☐ Define radioactive decay and radioisotope on SG pg. 201 ☐ Prepare an outline or narrative summary, write it on SG pg. 243

Supplies I Need for the Week

Things I Need to Prepare

Five Days a Week

Day 1	Day 2	Day 3	Day 4	Day 5
☐ Work on the research for the Scientist Biography Report, see SG pg. 239 for details	☐ Finish the research for the Scientist Biography Report	☐ Write an outline for the Scientist Biography Report, see SG pp. 240-242 for details	☐ Read pp. 128-129 from *UIDS*, then discuss what was read ☐ Write an outline on SG pg. 243	☐ Define radioactive decay and radioisotope on SG pg. 201 ☐ Enter the dates onto the date sheets on SG pp. 8-13

Supplies I Need for the Week

Things I Need to Prepare

Additional Information Week 34

Scientist Biography Report Step 2: Research the Scientist

☞ **From the Student Guide** – This week, you need to wrap-up the research you have done on your scientist. Read over your notes and make sure that you have at least five pieces of information for each of the categories below:

☑ Biographical information on the scientist (*i.e., where they were born, their parents, siblings, and how they grew up*);

☑ The scientist's education (*i.e., where they went to school, what kind of student they were, and what they studied*);

☑ Their scientific contributions (*i.e., research that they participated in, any significant discoveries they made, and the state of the world at the time of their contributions*).

Also, make sure that the recorded list of the resources you have used is complete.

Scientist Biography Report Step 3: Create an Outline

☞ **From the Student Guide** – Now that your research is completed, you are ready to begin the process of writing a report on your chosen scientist. This week, you are going to organize the notes you took during step two into a formal outline which you will use next week to write the rough draft of your report. Use the outline template provided on the student sheets as a guide. You should include information on why you chose the particular scientist in your introduction section. For the conclusion section of the outline, you need to include why you believe someone else should learn about your chosen scientist and your impression of the scientist (i.e., *Did you like the scientist? Do you feel that they made a significant impact on the field of chemistry?*).

☞ **Step 3 Notes** – The outline the students create can look like the one below.

Scientist Biography Outline

I. Introduction and Biological Information on the Scientist
(4-6 points)

II. The Scientist's Education
(4-6 points)

III. The Scientist's Contributions
(1-3 sub categories each with 4-5 points)

IV. Conclusion
(4-5 points)

Discussion Questions

1. What is radioactivity? (*UIDS pg. 128 - Radioactivity is the property of an unstable nuclei, where it tends to break up spontaneously into nuclei of other elements, emitting radiation in the process.*)

2. What are the three types particles emitted from a radioactive nucleus? (*UIDS pg. 128 - The three types of particles emitted from a radioactive nucleus are streams of alpha particles, streams of beta particles, and gamma rays.*)

3. What is disintegration, and when does it stop? (*UIDS pg. 128 - Disintegration is the splitting*

of an unstable nucleus into two parts, typically the nucleus and alpha or beta particle. It stops when the atom becomes stable.)

4. What is a Becquerel? (*UIDS pg. 129 - A Becquerel is a unit of radioactive decay that is equal to one nuclear disintegration per second.*)

5. What is a half-life? (*UIDS pg. 129 - A half-life is the time it takes for half the atoms in a radioactive substance to take the first decay step. It is also known as the rate of radioactive decay.*)

6. What is the basic difference between nuclear fission and fusion? (*UIDS pg. 129 - In fission, the atoms split, but in fusion the atoms come together.*)

Want More

⟳ **Half-Life** – Have the students do a fun activity about radioactive half-lives. You will need bite-sized candies or food, plus a timer. Give the students 32 pieces of the bite-sized food. After 2 minutes, have them eat 16 pieces. After 2 more minutes, have them eat 8 pieces. After 2 more minutes, have them eat 4 pieces. After 2 more minutes, have them eat 2 pieces. After 2 more minutes, have them eat 1 piece. After 2 more minutes, have them break the 1 piece in half and eat one of the halves. Contine the process until the students can't break the bits in half.

Sketch Week 34

There is no sketch this week to allow more time for you students to finish their scientist biography paper.

Student Assignment Sheet Week 35
Pollution

Scientist Biography Report Project

This week, you will complete steps four and five of the Scientist Biography Report Project. You will use your outline from last week to create a rough draft of your report. Then, you will edit and revise your paper to create the final draft. The instructions for this week's assignments are on the following Scientist Biography Report sheets.

Vocabulary & Memory Work

- ☐ Vocabulary: biodegradable, pollutant
- ☐ Memory Work—Review the elements of the periodic table that you have memorized this year.

Sketch

- ▨ There is no sketch this week.

Writing

- ᕙ Reading Assignment: *Usborne Illustrated Dictionary of Science* pp. 210 (Pollution)
- ᕙ Additional Research Readings:
  - 📖 Air Pollution: *USE* pg. 64, *KSE* pg. 453
  - 📖 Water Pollution: *USE* pg. 75, *KSE* pg. 452

Dates

- 🕐 No dates to be entered.

Schedules for Week 35
Two Days a Week

Day 1	Day 2
☐ Write the rough draft, revise and write the final draft of the Scientist Biography Report, see SG pp. 245-247 for details	☐ Read pg. 210 from *UIDS,* then discuss what was read ☐ Define biodegradable and pollutant on SG pg. 201 ☐ Prepare an outline or narrative summary, write it on SG pp. 248-249

Supplies I Need for the Week:

Things I Need to Prepare:

Five Days a Week

Day 1	Day 2	Day 3	Day 4	Day 5
☐ Write the rough draft of the Scientist Biography Report, see SG pg. 245 for details	☐ Read pg. 210 from *UIDS,* then discuss what was read ☐ Write an outline on SG pp. 248-249	☐ Revise the rough draft of the Scientist Biography Report ☐ Define biodegradable and pollutant on SG pg. 201	☐ Write the final draft of the Scientist Biography Report, see SG pp. 245-247 for details	☐ Take the Unit 7 Test

Supplies I Need for the Week:

Things I Need to Prepare:

Additional Information Week 35

Scientist Biography Report Step 4: Write a Rough Draft

☞ **From the Student Guide** – Last week, you created a formal outline for your scientist biography report; now, it is time to take that outline and turn it into a rough draft. Simply take the points on your outline, combine and add in sentence openers to create a cohesive paragraph. Here's what your rough draft should look like:

- ☑ Paragraph 1 (*from outline point I*): introduce the scientist;
- ☑ Paragraph 2 (*from outline point II*): tell about the scientist's education;
- ☑ Paragraph 3-5 (*from outline point III*): share the scientist's contributions (*one paragraph for each contribution*);
- ☑ Paragraph 6 (*from outline point IV*): share your thoughts on the scientist and why someone should learn about him or her.

You can choose to hand write or type up your rough draft on a separate sheet of paper. However, keep in mind that you will need a typed version for step five.

Scientist Biography Report Step 5: Revise to Create a Final Draft

☞ **From the Student Guide** – Now that you have a typed, double-spaced rough draft, look over it one more time to make any changes you would like. Then, have your teacher or one of your peers look over the paper for you to correct any errors and bring clarity to any of the unclear sections. Once this is complete, make the necessary changes to your paper to create your final draft. Print out your final paper and include it on the next page.

☞ **Step 5 Notes** – I have included a grading rubric for this assignment in the Appendix on pp. 269-270. If you would like to change things up for your student, have them create a poster or mini-book for the final draft of their Scientist Biography Report.

Discussion Questions

1. What is pollution? (*UIDS pg. 210 - Pollution is the release of undesirable substances into the natural environment that harm the natural process of the Earth.*)
2. What are the similarities and differences between acid rain and smog? (*UIDS pg. 210 - Both smog and acid rain are the results of pollution from the combustion of fuels. They also both have acidic properties that can damage buildings. However, smog is fog mixed with dust and soot, while acid rain is rainwater with a lower pH.*)
3. What is the Greenhouse effect? (*UIDS pg. 210 - The Greenhouse effect is when solar energy is trapped in the atmosphere by carbon dioxide, which causes a rise in the temperature.*)
4. What can cause ozone depletion? (*UIDS pg. 210 - Ozone depletion, a thinning of the ozone layer of gas in the upper atmosphere, is caused by the presence of chlorine in the atmosphere.*)
5. What causes eutrophication? (*UIDS pg. 210 - Eutrophication, an overgrowth of aquatic plants, is caused by an excess of fertilizers that have run off into rivers.*)

Want More

↻ **Effects of Acid Rain** – Have the students see the effects of acid rain by doing the "See for yourself" on *Usborne Science Encyclopedia* pg. 65.

Sketch Week 35

There is no sketch this week to allow more time for you students to finish their scientist biography paper.

Unit 7: Chemistry of Industry
Unit Test Answers

Vocabulary Matching

1. K
2. N
3. B
4. E
5. G
6. A
7. I
8. J
9. O
10. L
11. D
12. P
13. F
14. M
15. H
16. C

True or False

1. False (*Soap, which is one type of detergent, is prepared from fats.*)
2. True
3. False (*As alkanes and alkenes get longer, they change state from gas to liquid.*)
4. False (*Alkanes have a single carbon-to-carbon bond. Alkenes have a double carbon-to-carbon bond.*)
5. True
6. False (*Alkynes are organic compounds with a triple carbon-carbon bond.*)
7. True
8. False (*Petroleum is a mixture of alkanes that can be refined, while natural gas consists mainly of methane.*)
9. True
10. False (*Examples of natural polymers are rubber and starch. Examples of synthetic polymers are plastics and polyester.*)
11. False (*Limestone is added to a blast furnace to get rid of the impurities from the iron ore.*)
12. True
13. True
14. True
15. False (*The Greenhouse effect is when solar energy is trapped in the atmosphere by carbon dioxide, which causes a rise in the temperature.*)
16. True

Short Answer

1. Detergent molecules have a hydrophilic (water-loving) head with a functional group that makes the end polar. They also have a hydrophobic (grease-loving) tail made of a long chain hydrocarbon.
2. A substitution reaction is one where an atom or functional group is replaced by a different atom or functional group. An addition reaction is one where two molecules react to produce a single larger molecule.
3. The different homologous series have different functional groups, which give them different chemical properties.
4. The three main processes in refining are primary distillation, cracking, and reforming. In primary distillation, crude oil is heated and separated in a fractionating column. In cracking, the larger alkanes are broken into smaller alkanes and alkenes. In reforming, gasoline is produced by breaking up straight-chain alkanes and reassembling them as branched-chain molecules.

5. Polymers are substances made from repeating sequences of monomers bonded together.
6. Most pure metals are weak and soft, but an alloy contains a combination of metals, making the material stronger.
7. Disintegration is the splitting of an unstable nucleus into two parts, typically the nucleus and alpha or beta particle. It stops when the atom becomes stable.
8. Both smog and acid rain are the results of pollution from the combustion of fuels. They also both have acidic properties that can damage buildings. However, smog is fog mixed with dust and soot, while acid rain is rainwater with a lower pH.
9. Students should have filled out at least fifty of the element abbreviations below:

1	2	3	4	5	6	7	8	9	10	11	12	13	14	15	16	17	18
1 **H** Hydrogen 1.008																	2 **He** Helium 4.003
3 **Li** Lithium 6.941	4 **Be** Beryllium 9.012											5 **B** Boron 10.81	6 **C** Carbon 12.01	7 **N** Nitrogen 14.01	8 **O** Oxygen 16.00	9 **F** Fluorine 19.00	10 **Ne** Neon 20.18
11 **Na** Sodium 22.99	12 **Mg** Magnesium 24.31											13 **Al** Aluminum 26.98	14 **Si** Silicon 28.09	15 **P** Phosphorus 30.97	16 **S** Sulfur 32.07	17 **Cl** Chlorine 35.45	18 **Ar** Argon 39.95
19 **K** Potassium 39.10	20 **Ca** Calcium 40.08	21 **Sc** Scandium 44.96	22 **Ti** Titanium 47.87	23 **V** Vanadium 50.94	24 **Cr** Chromium 52.00	25 **Mn** Manganese 54.94	26 **Fe** Iron 55.85	27 **Co** Cobalt 58.93	28 **Ni** Nickel 58.69	29 **Cu** Copper 63.55	30 **Zn** Zinc 65.39	31 **Ga** Gallium 69.72	32 **Ge** Germanium 72.61	33 **As** Arsenic 74.92	34 **Se** Selenium 78.96	35 **Br** Bromine 79.90	36 **Kr** Krypton 83.80
37 **Rb** Rubidium 85.47	38 **Sr** Strontium 87.62	39 **Y** Yttrium 88.91	40 **Zr** Zirconium 91.22	41 **Nb** Niobium 92.91	42 **Mo** Molybdenum 95.94	43 **Tc** Technetium 98.91	44 **Ru** Ruthenium 101.1	45 **Rh** Rhodium 102.9	46 **Pd** Palladium 106.4	47 **Ag** Silver 107.9	48 **Cd** Cadmium 112.4	49 **In** Indium 114.8	50 **Sn** Tin 118.7	51 **Sb** Antimony 121.8	52 **Te** Tellurium 127.6	53 **I** Iodine 126.9	54 **Xe** Xenon 131.3
55 **Cs** Cesium 132.9	56 **Ba** Barium 137.3	*	72 **Hf** Hafnium 178.5	73 **Ta** Tantalum 181.0	74 **W** Tungsten 183.9	75 **Re** Rhenium 186.2	76 **Os** Osmium 190.2	77 **Ir** Iridium 192.2	78 **Pt** Platinum 195.1	79 **Au** Gold 197.0	80 **Hg** Mercury 200.6	81 **Tl** Thallium 204.4	82 **Pb** Lead 207.2	83 **Bi** Bismuth 209.0	84 **Po** Polonium [209]	85 **At** Astatine [210]	86 **Rn** Radon [222]
87 **Fr** Francium [223]	88 **Ra** Radium [226]	**	104 **Rf** Rutherfordium [261]	105 **Db** Dubnium [262]	106 **Sg** Seaborgium [266]	107 **Bh** Bohrium [264]	108 **Hs** Hassium [269]	109 **Mt** Meitnerium [268]	110 **Ds** Darmstadtium [272]	111 **Rg** Roentgenium [272]	112 **Cn** Copernicium [285]	113 **Nh** Nihonium [286]	114 **Fl** Flerovium [289]	115 **Mc** Moscovium [289]	116 **Lv** Livermorium [293]	117 **Ts** Tennessine [294]	118 **Og** Oganesson [294]

*Lanthanides

57	58	59	60	61	62	63	64	65	66	67	68	69	70	71
La Lanthanum 138.9	**Ce** Cerium 140.1	**Pr** Praseodymium 140.9	**Nd** Neodymium 144.2	**Pm** Promethium [145]	**Sm** Samarium 150.4	**Eu** Europium 152.0	**Gd** Gadolinium 157.3	**Tb** Terbium 158.9	**Dy** Dysprosium 162.5	**Ho** Holmium 164.9	**Er** Erbium 167.3	**Tm** Thulium 168.9	**Yb** Ytterbium 173.0	**Lu** Lutetium 175.0

**Actinides

89	90	91	92	93	94	95	96	97	98	99	100	101	102	103
Ac Actinium [227]	**Th** Thorium 232.0	**Pa** Protactinium 231.0	**U** Uranium 238.0	**Np** Neptunium [237]	**Pu** Plutonium [244]	**Am** Americium [243]	**Cm** Curium [247]	**Bk** Berkelium [247]	**Cf** Californium [251]	**Es** Einsteinium [252]	**Fm** Fermium [257]	**Md** Mendelevium [258]	**No** Nobelium [259]	**Lr** Lawrencium [262]

Unit 7: Chemistry of Industry
Unit Test

Vocabulary Matching:

1. Detergent ___

2. Soap ___

3. Alkane ___

4. Alkene ___

5. Ester ___

6. Homologous Series ___

7. Hydrocarbon ___

8. Natural Polymer ___

9. Synthetic Polymer ___

10. Alloy ___

11. Raw Materials ___

12. Slag ___

13. Radioactive Decay ___

14. Radioisotope ___

15. Biodegradable ___

16. Pollutant ___

A. A group of organic compounds that have the same general formula for all its members.

B. A class of saturated hydrocarbons.

C. A substance that was released into the atmosphere, oceans, or rivers that upset the natural processes of the Earth.

D. The basic materials used to make a product.

E. A class of unsaturated hydrocarbons with at least one carbon-to-carbon double bond.

F. The process by which a nucleus ejects particles by radiation until stability is reached.

G. A homologous series of organic compounds that give fruits and flowers their fragrances.

H. A term used to describe matter that can be broken down into simpler substances by bacteria.

I. A chemical compound that consists of only hydrogen and carbon.

J. A polymer that is produced by a living organism, such as cellulose, starch, chitin, and proteins.

K. A synthetic or organic substance, not prepared from a fat, that helps to remove grease and oil in water.

L. A mixture of two or more metals, or a metal, and a nonmetal.

M. An unstable nucleus that has a different number of neutrons than a stable nucleus.

N. A metallic salt derived from a fat that is used to help break up grease and oil so that it can be removed.

O. A polymer that is man-made, also known as plastics.

P. Impurities found in iron ore that are left over from the smelting process.

True or False

1. _____ All detergents are prepared from fats.

2. _____ Detergents work by lowering water's surface tension.

3. _____ As alkanes and alkenes get longer, they change state from liquid to gas.

4. _____ Alkanes have a double carbon-to-carbon bond. Alkenes have a single carbon-to-carbon bond.

5. _____ Esters are organic compounds that play a role in the smells of fruits and flowers.

6. _____ Alkynes are organic compounds with a single carbon-carbon bond.

7. _____ The five main products of primary distillation are refinery gas, gasoline, kerosene, diesel oil, and residue.

8. _____ Natural gas is a mixture of alkanes that can be refined, while petroleum consists mainly of methane.

9. _____ A homopolymer is made from identical monomers, while a copolymer is made from two or more different monomers

10. _____ Examples of synthetic polymers are rubber and starch. Examples of natural polymers are plastics and polyester.

11. _____ Coke is added to a blast furnace to get rid of the impurities from the iron ore.

12. _____ Steel is stronger than pure iron.

13. _____ The three types of particles emitted from a radioactive nucleus are streams of alpha particles, streams of beta particles, and gamma rays.

226

14. _____ A half-life is the time it takes for half the atoms in a radioactive substance to take the first decay step.

15. _____ Eutrophication is when solar energy is trapped in the atmosphere by carbon dioxide, which causes a rise in the temperature.

16. _____ Ozone depletion, a thinning of the ozone layer of gas in the upper atmosphere, is caused by the presence of chlorine in the atmosphere.

Short Answer

1. Describe a detergent molecule.

2. What is the difference between a substitution reaction and an addition reaction?

3. What is the difference between the different homologous series?

4. What are the three main processes in refining? Describe each.

5. Why are alloys better than a pure metal on its own?

6. What are polymers?

7. What is disintegration, and when does it stop?

8. What are the similarities and differences between acid rain and smog?

9. Fill in as many of the element abbreviations on the blank periodic table on the next page as you can. (Aim for at least fifty elements.)

Chemistry: Wrap-up

Year-end Test

Chemistry for the Logic Stage
Year-end Test Information & Answers

Year-end Test Information

The year-end test is on the vocabulary (*Vocabulary Matching section*), short answer questions from the tests (*Multiple Choice section*) and memory work (*Short Answer section*) throughout the year. You can choose to make it open notes or not. The purpose of this test is to help your students gain familiarity with the concept of a final exam, so that it won't be quite so overwhelming when they reach the high school years.

Year-end Test Answers

Vocabulary Matching

1. F	13. E	25. R
2. A	14. N	26. AB
3. D	15. J	27. AJ
4. G	16. Q	28. AD
5. I	17. P	29. Z
6. K	18. W	30. AF
7. C	19. U	31. AA
8. M	20. S	32. AH
9. O	21. Y	33. AG
10. L	22. X	34. AI
11. H	23. T	35. AC
12. B	24. V	36. AE

Multiple Choice

1. B	9. C	17. C
2. C	10. C,B,C,E,A	18. A
3. A,B,D	11. D	19. A,C,B
4. A, C	12. C,A,B	20. D
5. A,C,D	13. B	21. D
6. A	14. D	22. B
7. C	15. A	23. D
8. B	16. B	24. A,C

Short Answer

1. At constant temperature, the volume of a gas is inversely proportional to the pressure.
2. At constant pressure, the volume of a gas is directly proportional to the temperature.
3. A pure compound always contains the same elements in the same proportions.
4. In a chemical reaction, the mass of the products is equal to the mass of the reactants.
5. 6.02×10^{23}

6. If a change is made to a reaction in equilibrium, the reaction will adjust itself to counter the effects of that change.

Chemistry for the Logic Stage
Year-end Test

Vocabulary Matching

1. Electron Shell ___

2. Element ___

3. Compound ___

4. Atomic Number ___

5. Atomic Mass ___

6. Kinetic Theory ___

7. Sublimation ___

8. Density ___

9. Conductivity ___

10. Diffusion ___

11. Molecule ___

12. Mixture ___

13. Solubility ___

14. Solute ___

15. Solvent ___

16. Ionic Bonding ___

17. Covalent Bonding ___

A. A substance made up of one type of atom, which cannot be broken down by chemical reaction to form a simpler substance.

B. A combination of two or more elements or compounds that are not chemically combined.

C. Occurs when a substance changes directly from a solid to a gas without changing into a liquid.

D. A substance made up of two or more different elements that are chemically joined in fixed proportions.

E. The ability of a solute to be dissolved.

F. A region around the nucleus of an atom where a specific number of electrons can exist.

G. The number of protons in the nucleus of an atom.

H. A substance that is formed when two or more atoms chemically join together.

I. The average mass number of the atoms in a sample of an element.

J. The substance in which the solute dissolves to form a solution, typically a liquid.

K. The theory that states that as the temperature rises, particles move faster and therefore take up more space.

L. The spreading out of a gas to fill the available space of the container it is in.

M. A measure of the amount of matter in a substance compared to its volume.

N. The substance that dissolves in the solvent to form a solution.

O. The measure of a substance's ability to conduct heat or electricity.

P. A chemical bond between two atoms, in which each atom shares an electron.

Q. A strong chemical bond that is formed by the attraction between two ions of opposite charges.

18. Metallic Bonding ____

R. A type of compound that is formed when an acid and a base react.

19. Chemical Reaction ____

S. A substance which speeds up a chemical reaction without being changed by the reaction.

20. Catalyst ____

T. A compound that reacts with an acid to produce water and a salt.

21. Redox reaction ____

U. An interaction between the atoms of two substances where the atoms rearrange to form two new molecules.

22. Acid ____

V. A solution that resists changes in pH.

23. Base ____

W. A chemical bond where positive metal ions form a lattice structure with freely moving electrons between them.

24. Buffer ____

X. A hydrogen containing compound that splits in water to give hydrogen ions.

25. Salt ____

Y. A reaction that primarily involves the transfer of electrons between two substances.

26. Organic Compound ____

Z. A group of solid esters that are stored in living things as a source of reserve energy, also known as fats.

27. Polymer ____

AA. A natural polymer that is made up of glucose monomers.

28. Carbohydrates ____

AB. A compound that contains the element carbon.

29. Lipids ____

AC. The basic materials used to make a product.

30. Protein ____

AD. A group of organic compounds that consists of carbon, hydrogen and oxygen.

31. Starch ____

AE. A chemical compound that consists of only hydrogen and carbon.

32. Vitamins ____

AF. A natural polymer that is made up of amino acids.

33. Detergent ____

AG. A synthetic or organic substance, not prepared from a fat, that helps to remove grease and oil in water.

34. Alloy ____

AH. A group of organic compounds that are essential for the normal growth and nutrition of living things.

35. Raw Materials ____

AI. A mixture of two or more metals, or a metal and a nonmetal.

36. Hydrocarbon ____

AJ. A substance with a long-chain of molecules, formed from smaller molecules called monomers.

Chemistry Wrap-up ~ Year-end Test

Multiple Choice

1. What is the basic structure of an atom?

   A. Atoms have a nucleus composed of electrons and protons at the center and neutrons that fly around the nucleus in different shells or layers.

   B. Atoms have a nucleus composed of neutrons and protons at the center and electrons that fly around the nucleus in different shells or layers.

   C. Atoms have a nucleus composed of neutrons and electrons at the center and protons that fly around the nucleus in different shells or layers.

   D. None of the above

2. How are the elements on the periodic table organized?

   A. The elements in the periodic table are arranged according to their color.

   B. The elements in the periodic table are arranged in alphabetical order.

   C. The elements in the periodic table are arranged according to their atomic number.

   D. All of the above

3. Which of the following elements are metals?

   A. Iron

   B. Sodium

   C. Fluorine

   D. Copper

4. Which of the following elements are metalloids (or semi-metals)?

   A. Boron

   B. Tin

   C. Silicon

   D. Sulfur

5. Which of the following elements are nonmetals

   A. Nitrogen

   B. Iron

   C. Oxygen

   D. Carbon

6. What are the three states of matter typically found on Earth?

   A. The three states of matter are solid, liquid, and gas.

   B. The three states of matter are frozen, liquid, and steam.

   C. The three states of matter are solid, liquid, and mass.

   D. None of the above

7. What is surface tension?

   A. Surface tension is the ability for water to pull objects under the surface.

   B. Surface tension is what happens when molecules don't get along.

   C. Surface tension is the skin-like properties of a liquid's surface due to intermolecular forces.

   D. None of the above

8. What causes crystals to form a definite shape?

   A. Crystals form a definite shape because of the container they are in.

   B. Crystals form a definite shape because of the arrangement of their atoms or ions.

   C. Crystals form a definite shape because of the location of attachment.

   D. All of the above

9. What is the difference between concentrated and dilute solutions?

   A. Concentrated solutions have the same amount of solute in relation to solvent. Dilute solutions have a small amount of solute in relation to solvent.

Chemistry Wrap-up ~ Year-end Test

B. Concentrated solutions have a small amount of solute in relation to solvent. Dilute solutions have a large amount of solute in relation to solvent.

C. Concentrated solutions have a large amount of solute in relation to solvent. Dilute solutions have a small amount of solute in relation to solvent.

D. None of the above

10. Match the separation technique with what it is typically used to do.

Evaporation _____

A. Is used to separate two or more liquids when the liquids have similar boiling points.

Filtration _____

B. Is used to separate a suspension of particles of a solid that are spread through a liquid.

Centrifuging _____

C. Is a way to separate a soluble solid that has been dissolved in a liquid solvent.

Distillation _____

D. Is used to separate a suspension when filtering or settling will not work.

Fractional distillation _____

E. Is used to separate a mixture of two or more liquids.

11. What types of analysis do chemists use?

A. Qualitative analysis

B. Quantitative analysis

C. Structural analysis

D. All of the above

12. Match the type of bonding to its definition.

Ionic Bonding _____

A. A chemical bond between two atoms, in which each atom shares an electron.

Covalent Bonding _____

B. A chemical bond where positive metal ions create a lattice structure with freely moving electrons between them.

Metallic Bonding _____

C. A strong chemical bond that is caused by the attraction between two ions of opposite charges.

13. What is the difference between exothermic and endothermic reactions?

    A. Endothermic reactions release heat; exothermic reactions require heat.

    B. Endothermic reactions require heat; exothermic reactions release heat.

    C. None of the above.

14. What does the reactivity series tell us?

    A. The reactivity series tells us how reactive certain metals are.

    B. The reactivity series tells us that metals at the top of the series are very reactive.

    C. The reactivity series tells us that the metals at the bottom of the series are the least reactive.

    D. All of the above.

15. What happens to a substance when it is reduced?

    A. When a substance is reduced, it loses oxygen or gains hydrogen in a chemical reaction.

    B. When a substance is reduced, it gains oxygen or loses hydrogen in a chemical reaction.

    C. None of the above.

16. What happens to a substance when it is oxidized?

    A. When a substance is oxidized, it loses oxygen or gains hydrogen in a chemical reaction.

    B. When a substance is oxidized, it gains oxygen or loses hydrogen in a chemical reaction.

    C. None of the above.

17. What happens to acids in water?

    A. Acids cannot be dissolved in water, so nothing happens.

    B. Acids and water do not mix, so the acid floats on top of the water.

    C. When acids dissolve in water they yield positively charged hydrogen ions.

    D. Acids dissolve in water to form gas and a base.

18. What do bases do to acids?

    A. Bases neutralize acids by accepting hydrogen ions.

    B. Bases make acids stronger by giving away hydrogen ions.

    C. Bases do not affect acids.

    D. None of the above.

19. Match the term with the pH scale values.

    Acidic _____   A. pH 0 to 6.9

    Neutral _____   B. pH 7.1 to 14

    Basic _____   C. pH 7

20. Which of the groups of chemicals below are essential to life?

    A. Proteins

    B. Carbohydrates

    C. Vitamins

    D. All of the above.

21. What are enzymes and how does our body use them?

    A. Enzymes are biological catalysts.

    B. Enzymes are used by our body to break down our food.

    C. Enzymes are man-made and only used in the lab.

    D. Both A and B.

22. Why is yeast used in alcohol fermentation?

    A. Yeast is used because it slows the reaction down.

    B. Yeast in used in alcohol fermentation because it has the enzyme zymase.

    C. None of the above

23. What is special about soaps and detergents?

    A. Soaps and detergents have a water-loving head and a grease-loving tail.

    B. Soaps and detergents are able to dissolve in water.

    C. Soaps and detergents can attract grease and remove it from a surface.

    D. All of the above.

24. What is true about homologous series of organic compounds?

    A. Homologous series have different functional groups.

    B. Homologous series have the same chemical formula.

    C. Homologous series have different chemical properties.

    D. None of the above.

Short Answer

1. What is Boyle's Gas Law?

2. What is Charles' Gas Law?

3. What is the Law of Constant Composition?

4. What is the Law of Conservation of Mass?

5. What is Avogadro's number?

6. What does Le Chatelier's Principle say?

Appendix

Ancients 5000 BC–400 AD

- 🕐 4000 BC – Metals like lead and silver are melted down from ores and used.
- 🕐 c. 3000 BC – The first alloy is created during the Bronze Age.
- 🕐 3000 BC – Yeast is used to make alcoholic drinks, such as beer or wine through fermentation.
- 🕐 340 BC – Aristotle proposes that all substances are made up of combinations of four elements: earth, air, water and fire.
- 🕐 287 BC-212 BC – Archimedes lived. He used the principle of density to determine if King Hiero II's crown was made of real gold.
- 🕐 200 – Yogurt is made through the use of bacteria.
- 🕐 347 – Oil wells are first drilled in China. The oil was naturally refined and used as a building adhesive, for water-proofing and as a fuel.

Medieval–Early Renaissance 400AD–1600AD

- 🕐 11th century – Arabic chemists discover how to make acidic compounds such as sulfuric, nitric and hydrochloric acids.

Late Renaissance–Early Modern 1600 AD–1850 AD

- 🕐 1620 – Francis Bacon publishes a book, *New Method*, in which he states that theories need to be supported by experiments.
- 🕐 1649 – Pierre Gassendi states that atoms can be joined together to form molecules.
- 🕐 1661 – Robert Boyle publishes *The Sceptical Chymist*, in which he says ideas should always be tested through experiments.
- 🕐 1675 – Robert Boyle, an Irish chemist, suggests that acids might be made up of sharp pointed particles.

Late Renaissance-Early Modern 1600 AD-1850 AD

- 1704 – Isaac Newton roughly outlines atomic theory.
- 1722 – Antoine Lavoisier shows that diamonds are a form of carbon by burning samples of charcoal and diamond. He finds that neither produced any water and that both released the same amount of carbon dioxide per gram.
- 1754 – French chemist, Guillaume-François Rouelle, states that a base is a substance that reacts with an acid to produce a solid (or salt).
- 1766-1844 – John Dalton lives. He is responsible for writing the first atomic theory.
- 1767 – Joseph Priestley inventes carbonated water when he discovers that he can infuse water with carbon dioxide by suspending a bowl of water above a beer.
- 1770's – Antoine Lavoisier, a French chemist, suggests that acids form oxygen compounds when dissolved in water, so they must all contain oxygen.
- 1772 & 1774 – Swedish chemist Carl Scheele and English chemist Joseph Priestley both discover oxygen.
- 1778-1829 – English chemist Humphrey Davy lives. He is one of the first scientists to see electrolysis in action.
- 1789 – Antoine Lavoisier establishes that matter cannot be created or destroyed.
- 1798-1808 – Joseph-Louis Proust analyzes the different sources of several compounds and finds that their elements always contained the same ratio by weight. This leads to the discovery of the law of constant composition.
- 1800 – William Nicholson and Johann Ritter decomposed water into hydrogen and oxygen.
- 1808 – Jons Berzelius, a Swedish chemist, first uses the term organic chemistry to refer to the chemistry of living things.
- 1811 – Italian scientist Amedeo Avogadro first proposes that the volume of a gas at a constant pressure and temperature is proportional to the number of atoms or molecules regardless of the nature of the gas.
- 1813 – The first book on food chemistry, entitled *Elements of Agricultural Chemistry*, is published by Sir Humphrey Davy.
- 1827 – Robert Brown observes pollen grains randomly bouncing round in water, discovering Brownian Motions, which happen because the grains are being constantly hit by water molecules.
- 1828 – The meaning of organic chemistry changes to refer to the chemistry of carbon when Friedrich Wohler succeeds in synthesizing a natural carbon compound in the lab.
- 1832 – Michael Faraday writes a mathematical equation that can be used to calculate the quantity of the separated elements in electrolysis.
- 1833 – The first enzyme diastase, which we call amylase today, is discovered by Anselme Payen.

Modern 1850 AD-Present

- 🕐 1850's – Alexander Parkes, an English chemist, synthesizes the first polymer.
- 🕐 1850's – The first commercial oil refinery appears.
- 🕐 1850-1936 – Henry Louis Le Chatelier lives. He is famous for the equilibrium law that he discovers during these years.
- 🕐 1856 – Henry Bessemer invents a way of removing most of the carbon from iron, creating steel.
- 🕐 1860 – Ludwig Boltzmann develops the kinetic theory of gases, which is widely opposed by other scientists.
- 🕐 1869 – Russian chemist Dmitri Mendeleyev draws up the very first periodic table, leaving gaps for elements that had not yet been discovered.
- 🕐 1887 – Svante Arrhenius, a Swedish chemist, develops the modern theory on acids. He states that all acids contain hydrogen ions, which give these compounds their unique properties.
- 🕐 1896 – Antoine Becquerel is working with a natural fluorescent material and x-rays, which leads to his discovery of radioactivity.
- 🕐 1897 – J.J. Thompson discovers the electron.
- 🕐 1897 – Sir William Crookes discovers a fourth state of matter through his experiments with gases.
- 🕐 1898 – Marie and Pierre Curie coin the term "radioactive" when they discovered radium and polonium.
- 🕐 1909 – Danish chemist Soren Sorenson comes up with a logarithmic scale to show the concentration of the hydrogen ions in a solution, which ranged from 0 to 14. Today this scale is known as the pH scale.
- 🕐 1909 – Rutherford, along with two other scientists, discovers the nucleus of an atom.
- 🕐 1909 – Wilhelm Ostwald is awarded the Nobel Prize for Chemistry for his work with catalysts.
- 🕐 1913 – Niels Bohr comes up with the Bohr model of an atom.
- 🕐 1915 – William Bragg and his son receive the Nobel Prize in Physics for their work on crystalline structures.
- 🕐 1916 – Gilbert Lewis develops the concept of electron pair bonding.
- 🕐 1919 – English chemist Francis Ashton invents the mass spectrometer.
- 🕐 1923 – Danish chemist Johannes Bronsted and English chemist Martin Lowry both suggest a change to Arrhenius' definition of acids and bases. They define an acid as a chemical that donates hydrogen ions (or protons) and a base as one that accepts them.
- 🕐 1926 – Jean Perrin wins the Nobel Prize in Physics for determining the constant that Avogadro first proposed. He named the constant Avogadro's number in honor of the scientist.
- 🕐 1928 – US scientist Irving Langmuir names the fourth state of matter plasma.
- 🕐 1928 – Hans Geiger and one of his students develop a machine that can detect and measure the intensity of radiation.
- 🕐 1929 – Linus Pauling sets forth a set of rules called "Pauling's Rules", which we still use today when rationalizing the crystalline structure of ionic solids.

Modern 1850 AD-Present

- 1950 to today – The field of biochemistry, the study of the chemistry within living things, has greatly advanced with the development of new technology such as chromatography, X-ray diffraction, NMR spectroscopy, radio-isotopic labeling, electron microscopy, and molecular dynamics' simulations.
- 1954 – Linus Pauling is awarded the Nobel Prize for Chemistry for his work on calculating the energy to break bonds.
- 1954 – Morris Tanenbaum invents the first silicon transistor at Bell Labs, beginning the use of this semiconductor in electronics.
- 1965 – Stephanie Kwolek, an American chemist for DuPont, creates the first Kevlar fibers, a polymer that is used to protect law enforcement officers from bullets.

The Scientific Method Explained

The scientific method is a method for asking and answering scientific questions. This is done through observation and experimentation.

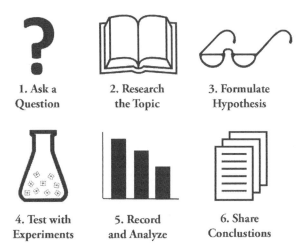

The following steps are key to the scientific method:

1. **Ask A Question** – The scientific method begins with asking a question about something you observe. Your questions must be about something you can measure. Good questions begin with how, what, when, who, which, why, or where.

2. **Do Some Research** – You need to read about the topic from your question so that you can have background knowledge of the topic. This will keep you from repeating mistakes made in the past.

3. **Formulate a Hypothesis** – A hypothesis is an educated guess about the answer to your question. Your hypothesis must be easy to measure and answer the original question you asked.

4. **Test with Experimentation** – Your experiment tests whether your hypothesis is true or false. It is important for your test to be fair. This means that you may need to run multiple tests. If you do, be sure to only change one factor at a time so that you can determine which factor is causing the difference.

5. **Record and Analyze Observations or Results** – Once your experiment is complete, you will collect and measure all your data to see if your hypothesis is true or false. Scientists often find that their hypothesis was false. If this is the case, they will formulate a new hypothesis and begin the process again until they are able to answer their question.

6. **Draw a Conclusion** – Once you have analyzed your results, you can make a statement about them. This statement communicates your results to others.

Reading Assignments for Younger Students

Unit 1: The Periodic Table

Week	Topic Studied	Resource and Pages Assigned
Week 1	Atoms	*DK Smithsonian Science: A Visual Encyclopedia pp. 10-11 (Atoms)*
Week 2	Periodic Table	*DK Smithsonian Science: A Visual Encyclopedia pp. 24-25 (All the Elements)*
Week 3	Metals	*DK Smithsonian Science: A Visual Encyclopedia pp. 36-37 (Metals)*
Week 4	The Inbetweens	*DK Smithsonian Science: A Visual Encyclopedia pp. 38-39 (Strange Metals)*
Week 5	Halogens and Noble Gases	*DK Smithsonian Science: A Visual Encyclopedia pp. 40-41 (Nonmetals)*

Unit 2: Matter

Week	Topic Studied	Resource and Pages Assigned
Week 6	States of Matter	*DK Smithsonian Science: A Visual Encyclopedia pp. 8-9 (Defining Matter), pp. 22-23 (Changing States)*
Week 7	Solid Structures	*DK Smithsonian Science: A Visual Encyclopedia pp. 14-15 (Solids)*
Week 8	Molecular Properties	*DK Smithsonian Science: A Visual Encyclopedia pp. 18-19 (Liquids)*
Week 9	Gas Laws	*DK Smithsonian Science: A Visual Encyclopedia pp. 10-11 (Gases Galore)*

Unit 3: Solutions

Week	Topic Studied	Resource and Pages Assigned
Week 10	Compounds and Mixtures	*DK Smithsonian Science: A Visual Encyclopedia pp. 12-13 (Molecules)*
Week 11	Solutions	*DK Smithsonian Science: A Visual Encyclopedia pp. 28-29 (Solutions and Solvents)*
Week 12	Separating Mixtures and Compounds	*DK Smithsonian Science: A Visual Encyclopedia pp. 26-27 (Mixtures)*
Week 13	Electrolysis	*No DK pages scheduled, read suggested pages from the Usborne Science Encyclopedia*

Reading Assignments for Younger Students

Unit 4: Chemical Reactions

Week	Topic Studied	Resource and Pages Assigned
Week 14	Chemical Bonding	*No DK pages scheduled, read suggested pages from the Usborne Science Encyclopedia*
Week 15	Chemical Reactions	*DK Smithsonian Science: A Visual Encyclopedia pp. 32-33 (Incredible Reactions)*
Week 16	Reactivity	*No DK pages scheduled, read suggested pages from the Usborne Science Encyclopedia*
Week 17	Chemical Reactions	
Week 18	Catalysts	
Week 19	Oxidation and Reduction	

Unit 5: Acids and Bases

Week	Topic Studied	Resource and Pages Assigned
Week 20	Acids	*DK Smithsonian Science: A Visual Encyclopedia pg. 30 (Acids)*
Week 21	Bases	*DK Smithsonian Science: A Visual Encyclopedia pg. 31 (Bases)*
Week 22	Measuring Acidity (pH)	*No DK pages scheduled, read suggested pages from the Usborne Science Encyclopedia*
Week 23	Neutralization and Salts	

Unit 6: Chemistry of Life

Week	Topic Studied	Resource and Pages Assigned
Week 24	Organic Chemistry	*DK Smithsonian Science: A Visual Encyclopedia pp. 56-57 (Organic Chemistry)*
Week 25	Enzymes	*No DK pages scheduled, read suggested pages from the Usborne Science Encyclopedia*
Week 26	Chemistry of Food	
Week 27	Alcohols	

Unit 7: Chemistry of Industry

Week	Topic Studied	Resource and Pages Assigned
Week 28	Soaps and Detergents	*No DK pages scheduled, read suggested pages from the Usborne Science Encyclopedia*

Reading Assignments for Younger Students

Unit 7: Chemistry of Industry, continued

Week 29	Alkanes and Alkenes	*No DK pages scheduled, read suggested pages from the Usborne Science Encyclopedia*
Week 30	Homologous Groups	
Week 31	Petrochemicals	*DK Smithsonian Science: A Visual Encyclopedia pp. 76-77 (Industrial Materials)*
Week 32	Polymers and Plastics	*DK Smithsonian Science: A Visual Encyclopedia pp. 62-63 (Plastics)*
Week 33	Iron and Alloys	*No DK pages scheduled, read suggested pages from the Usborne Science Encyclopedia*
Week 34	Radioactivity	*DK Smithsonian Science: A Visual Encyclopedia pp. 136-137 (Radioactivity and Nuclear Power)*
Week 35	Pollution	*DK Smithsonian Science: A Visual Encyclopedia pp. 78-79 (Recycling)*

Element Trading Cards
Card Directions and Games

Directions for Making the Cards

Have the students make a 3x5 card for several or all of the elements. Each element card should include the element name and abbreviation, its atomic number and mass, who discovered the element and when it was discovered, along with some common uses of the element. The students can choose to write out the uses or illustrate them.

Sample Card

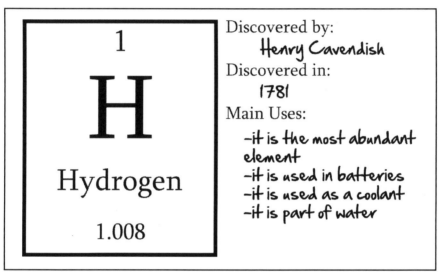

Games for the Element Trading Cards

- **Guess the Element** – Set the deck of element cards face down in the center of the players. Have the first player choose a card and begin to read the information on the card, i.e., "This element is 8th on the periodic table; this element is essential to life; it is in the air; it was discovered by Joseph Priestly." and so on. The other players try to call out what the element is as soon as they figure it out. The first person to guess which element it is wins a point. If no one guesses the element, then no one gets the point. The first person to reach 10 points is the winner of the game.

- **Periodic Table Match-up** – You will need a set of element cards, a copy of the periodic table and a pen for each student. Choose one player to be the caller. Have them choose an element from the deck of cards and call out its name. The caller will begin counting to 12, if the players find the element within 6 seconds, they put a check on the element. If they find it within 12 seconds, they will circle the element. Once the caller reaches 12; they will give the atomic number of the element and the players who haven't found the element yet will put an X over it. Checks are worth 2 points, circles are worth 1 point and X's are worth no points. The game ends once all of the elements have been called. The player with the highest points wins the game.

Element Trading Cards
Card Templates

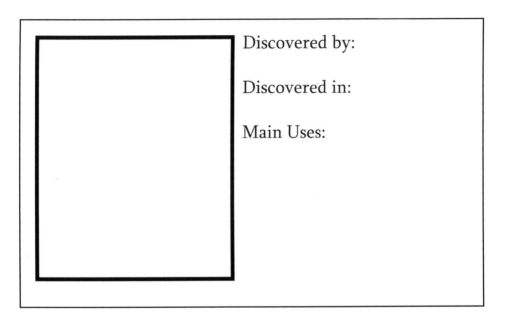

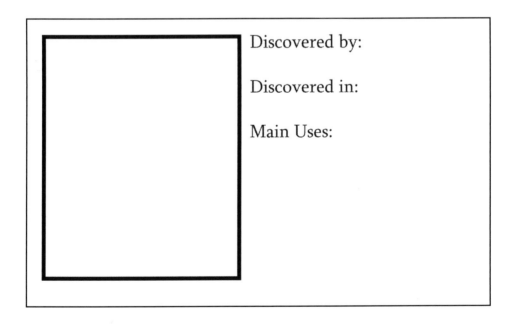

Electrical Circuit Diagram and Instructions

The following electrical circuit will be needed for the week four experiment.

Supplies Needed:
- ✓ Battery (D-Cell or larger)
- ✓ Wire (at least 3 feet)
- ✓ Flashlight bulb
- ✓ 2 Alligator clips
- ✓ Electrical Tape

Instructions:
1. Cut the wire into three lengths.
2. Take one of the wires and attach an alligator clip to one end.
3. Take the next wire and attach one end to the other alligator clip. Then, wrap the other end once around the bottom of a flashlight bulb. Tape a bit of electrical tape around the wire near the bulb so that you can handle it.
4. Now take the wire with bare end and it to one of the terminals of the battery. Then, you can tap the end of the bulb to the other terminal of the battery to test the circuit. (*See the pictures below to visually check your electrical circuit.*)

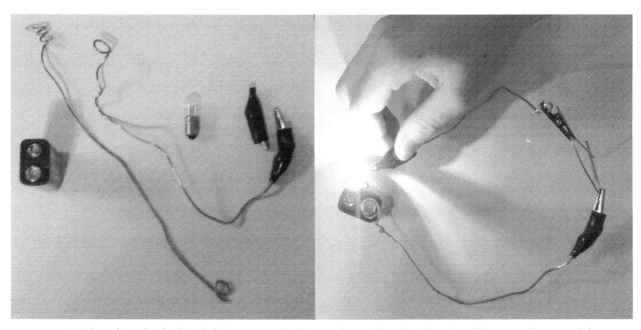

5. The electrical circuit is now ready for testing. Simply clip the alligator clips on either side of your sample to complete the circuit. If the bulb lights up, the material conducts electricity; if it does not, the sample does not conduct electricity.

The Science Fair Project Presentation Board

The science fair project board is the visual representation of the students' hard work. Below is a list of the information that needs to be included along with where it is typically found on the project board. The students can certainly mix things up a bit, but be sure to remind them that their information needs to be placed in such a way that it is easy for someone else to follow.

The left section of the board typically has:
- ✓ Introduction
- ✓ Hypothesis
- ✓ Research

The center section of the board typically has:
- ✓ Materials
- ✓ Procedure
- ✓ Pictures, Graphs, and Charts from the Experiment (*Note—The students can also display a portion of their project or a photo album with pictures from their experiment on the table in front of their board.*)

The right section of the board typically has:
- ✓ Results
- ✓ Conclusions

When purchasing a presentation board, you are looking for a tri-fold board that is at least 36" high and 48" wide, with the two side sections folding into the center section. If you search the internet for science fair project boards, you will find plenty of options, some with header boards, some without. You can usually purchase a project board at most local WalMart, Target or Michael's stores.

Here is a description of what the students need to prepare for their presentation board.

- ↳ **The Introduction** – *Have the students turn their questions from step one into a statement. Then, they should write two to three more sentences explaining why they chose their specific topics. The students should end their introductory paragraphs by sharing the question that they were trying to answer with their project.*

- ↳ **The Hypothesis** – *Have the students type up and prepare their hypotheses from step three for the project board.*

- ↳ **The Research** – *Have the students type up and prepare their research reports from step two for the project board.*

- ↳ **The Materials** – *Have the students type up a list of the materials they used for their projects.*

- ↳ **The Procedure** – *Have the students revise the experiment design they wrote in step four so that it is written in the past tense.*

- ↳ **The Results** – *Have the students turn the trends in the observations they noted and the results they interpreted in step six into a paragraph.*

- ↳ **The Conclusion** – *Have the students type up and prepare their concluding paragraphs from step six for the project board.*

Chemistry for the Logic Stage ~ Appendix

Density Calculations

Density is a property of matter that relates mass and volume. Remember that mass tells you how much matter is in something, while volume tells you how much space something occupies. So, in other words, density measures how much matter of a particular substance will occupy a given space. We use the equation below to calculate the density of a solution:

$$\text{Density} = \frac{(\text{mass of water} + \text{mass of salt})}{\text{volume of water}}$$

Sample calculation

So, let's try to calculate the density of a solution that contains ¾ cup of salt in 2 cups of water. We know that 1 TBSP of salt has a mass of 17.1 g, and that there are 4 TBSP in ¼ of a cup. We also know that 1 cup of water has an approximate volume of 237 mL and a mass of 237 g. We can use what we know with the formula above to calculate the density of the solution. (*Note—You do not need to take into account volume for the salt because it is dissolved in the water, meaning it fits into the spaces within the cup of water, so that the volume change is negligible.*)

$$\text{Density} = \frac{(2 \times 237)\text{g} + (3 \times (4 \times 17.1))\text{g}}{(2 \times 237)\text{mL}}$$

$$\text{Density} = \frac{474\text{g} + 205.2\text{g}}{474\text{mL}} = \frac{679.2\text{g}}{474\text{mL}}$$

$$\text{Density} = 1.43 \ \frac{\text{g}}{\text{mL}}$$

Now, it is your turn to calculate the densities of the solutions you made for the week seven experiment.

A Quick Word about Significant Figures

In chemistry, we use significant figures when we round our answers. Significant figures show us the certainty to which we are confident with an answer. So, if the numbers in my calculations all contain three digits, my answer should contain three digits. There are several rules that govern whether or not a number is significant that you will learn later on in your studies. For now, just remember that all your densities should be rounded to show three digits, just like in the sample calculation.

258

Week 8 Experiment Density Calculations

Density of the yellow solution (no salt)

$$\text{Density} = \frac{+}{}$$

Density =

Density of the red solution (2 TBSP of salt)

$$\text{Density} = \frac{+}{}$$

Density =

Density of the blue solution (¼ cup of salt)

$$\text{Density} = \frac{+}{}$$

Density =

Density of the green solution (6 TBSP of salt)

$$\text{Density} = \frac{+}{}$$

Density =

Nomenclature Worksheet

Nomenclature, or naming compounds in chemistry, came about when scientist wanted a consistent way to refer to the compounds they were discovering. There are several basic things we need to know before we can learn how to name compounds in chemistry.

First, we use prefixes to distinguish compounds that have the same two elements in differing numbers. The first six prefixes, which are derived from Latin, are *mono* ("one"), *di* ("two"), *tri* ("three"), *tetra* ("four"), *penta* ("five"), and *hexa* ("six"). The prefix *mono*, is usually left out from the beginning of the first word of the name.

Second, many compounds have common names which we use instead. Over the years, it has become traditionally accepted to use these names over their systematic names. For example:

☞ C_2H_6 - Ethane instead of dicarbon hexahydride;

☞ H_2O - Water instead of dihydrogen oxide;

☞ NH_3 - Ammonia instead of nitrogen trihydride.

Finally, polyatomic ions, or ions with more than one atom, have special names. For example:

☞ SO_4^{-2} - Sulfate

☞ OH^- - Hydroxide

☞ NO_3^- - Nitrate

☞ CO_3^{-2} - Carbonate

☞ PO_4^{-3} - Phosphate

When naming compounds, we use a different set of rules to name ionic and covalent compounds. Here are the rules to follow:

Ionic Compounds
(metal and nonmetal)
1. Write the name for the cation, which is the metal ion.
2. Write the name for the anion, which is the nonmetal ion or polyatomic ion. (**Note**—*If the anion is a nonmetal atom, use the root name and add the suffix –ide.*)

Examples
☞ NaCl – Sodium chloride
☞ HCl - Hydrogen chloride
☞ $CaSO_4$ – Calcium sulfate

Covalent Compounds
(two nonmetals)
1. Write the name of the first element. If the symbol is followed by a subscript of two or more, use the proper prefix to show the number.
2. Write the root name of the second element and add the suffix –ide. If the symbol is followed by a subscript of two or more, use the proper prefix to show the number.

Examples
☞ CO - Carbon monoxide
☞ N_2O_4 - Dinitrogen tetroxide
☞ PCl_3 – Phosphorus trichloride

Nomenclature Practice

Ionic Compounds

MgO _____

$NaBr$ _____

Li_2S _____

$MgSO_4$ _____

$Be(OH)_2$ _____

$Sr(NO_3)_2$ _____

Covalent Compounds

CO_2 _____

NO_2 _____

SO_3 _____

N_2S _____

BF_3 _____

P_2Br_4 _____

Chemical Equations Worksheet

How to Balance a Chemical Equation – Part 1

When we balance a chemical equation, we are making sure that the elements involved are present in equal amounts on both sides of the equation. Let's look at the equation below:

$$H_2 + O_2 \longrightarrow H_2O$$
(Hydrogen)   (Oxygen)   (Water)

We have:

Elements	Reactants	Products
Hydrogen (H)	2	2
Oxygen (O)	2	1

As you can see there are 2 hydrogen atoms on the reactants' side of the equation and 2 hydrogen atoms on the products' side of the reaction. However, there are 2 oxygen atoms on the reactants' side of the equation and only 1 oxygen atom on the products' side of the reaction. This makes the equation unbalanced. We can not change the subscripts in the molecule or it will change the molecule itself, so we need to add a coefficient to balance the equation. A coefficient is the big number in front of the element symbol that tells us how many molecules there are in the equation.

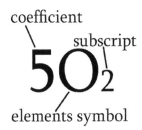

So, what if we added a coefficient of 2 before each of the molecules involved? This would give us:

$$2H_2 + 2O_2 \longrightarrow 2H_2O$$
(Hydrogen)   (Oxygen)   (Water)

Now we have:

Elements	Reactants	Products
Hydrogen (H)	4	4
Oxygen (O)	4	2

The equation is still not balanced. What if we added a coefficient of 2 only to the hydrogen and water molecules?

$$2H_2 + O_2 \longrightarrow 2H_2O$$
(Hydrogen)   (Oxygen)   (Water)

Now we have:

Elements	Reactants	Products
Hydrogen (H)	4	4
Oxygen (O)	2	2

Our equation is now balanced.

If you noticed in the equation above, we didn't write a coefficient if we only had one molecule. This is because we don't normally write a coefficient of 1; instead, the absence of a coefficient implies that there is only one of that molecule needed for the reaction. Now it's your turn to try balancing a few chemical equations.

Balancing Chemical Equations – Part 1 Practice

Add the necessary coefficients to balance the following equations. Use the charts to help you determine which coefficients you need to use.

1. ____ $AgNO_3$ + ____ Cu ⟶ ____ $Cu(NO_3)_2$ + ____ Ag

Elements	Reactants start	Products start	Reactants final	Products final
Silver (Ag)	1	1		
Nitrogen (N)	1	2		
Oxygen (O)	3	6		
Copper (Cu)	1	1		

2. ____ $AlBr_3$ + ____ K ⟶ ____ KBr + ____ Al

Elements	Reactants start	Products start	Reactants final	Products final
Aluminum (Al)				
Bromine (Br)				
Potassium (K)				

3. ____ $LiNO_3$ + ____ $CaBr_2$ ⟶ ____ $Ca(NO_3)_2$ + ____ $LiBr$

Elements	Reactants start	Products start	Reactants final	Products final
Lithium (Li)				
Nitrogen (N)				
Oxygen (O)				
Calcium (Ca)				
Bromine (Br)				

4. ____ $PbBr_2$ + ____ HCl ⟶ ____ HBr + ____ $PbCl_2$

Elements	Reactants start	Products start	Reactants final	Products final
Lead (Pb)				
Bromine (Br)				
Hydrogen (H)				
Chlorine (Cl)				

5. ____ $NaCN$ + ____ $CuCO_3$ ⟶ ____ Na_2CO_3 + ____ $Cu(CN)_2$

Elements	Reactants start	Products start	Reactants final	Products final
Sodium (Na)				
Carbon (C)				
Nitrogen (N)				
Copper (Cu)				
Oxygen (O)				

Balancing a Chemical Equations for Reversible Reactions – Part 2

Review Practice – Balance the following chemical equation.

1. ____ Al + ____ HCl ⟶ ____ H_2 + ____ $AlCl_3$

Elements	Reactants start	Products start	Reactants final	Products final
Aluminum (Al)				
Hydrogen (H)				
Chlorine (Cl)				

Balancing Chemical Equations for Reversible Reactions

Reversible reactions are also known as equilibrium reactions. The equations for these reactions depict two simultaneous reactions, known as the forward reaction and the reverse reaction. When balancing these equations, you will follow the same procedure as before.

Practice – Add the necessary coefficients to balance the following equations. Use the charts to help you determine which coefficients you need to use. (**Note**—*When completing the charts for balancing these reactions, use the forward reaction, meaning that the left side of the equation contains the reactants and the right side of the equation contains the products.*)

2. ____ N_2 + ____ O_2 ⇌ ____ NO_2

Elements	Reactants start	Products start	Reactants final	Products final
Nitrogen (N)				
Oxygen (O)				

3. ____ NO_2 ⇌ ____ N_2O_4

Elements	Reactants start	Products start	Reactants final	Products final
Nitrogen (N)				
Oxygen (O)				

4. ____ SO_3 ⇌ ____ SO_2 + ____ O_2

Elements	Reactants start	Products start	Reactants final	Products final
Sulfur (S)				
Oxygen (O)				

5. ____ N_2 + ____ H_2 ⇌ ____ NH_3

Elements	Reactants start	Products start	Reactants final	Products final
Nitrogen (N)				
Hydrogen (H)				

Review Practice for Balancing a Chemical Equations – Part 3

Add the necessary coefficients to balance the following equations. Use the charts to help you determine which coefficients you need to use.

1. ____ FeO + ____ PdF_2 ⟶ ____ FeF_2 + ____ PdO

Elements	Reactants start	Products start	Reactants final	Products final
Iron (Fe)				
Oxygen (O)				
Palladium (Pd)				
Fluorine (F)				

2. ____ $Si(OH)_4$ + ____ NaBr ⟶ ____ $SiBr_4$ + ____ NaOH

Elements	Reactants start	Products start	Reactants final	Products final
Silicone (Si)				
Oxygen (O)				
Hydrogen (H)				
Sodium (Na)				
Bromine (Br)				

3. ____ $RbNO_3$ + ____ BeF_2 ⟶ ____ $Be(NO_3)_2$ + ____ RbF

Elements	Reactants start	Products start	Reactants final	Products final
Rubidium (Rb)				
Nitrogen (N)				
Oxygen (O)				
Beryllium (Be)				
Fluorine (F)				

4. ____ N_2 + ____ F_2 ⟶ ____ NF_3

Elements	Reactants start	Products start	Reactants final	Products final
Nitrogen (N)				
Fluorine (F)				

5. ____ H_2 + ____ I_2 ⇌ ____ HI

Elements	Reactants start	Products start	Reactants final	Products final
Hydrogen (H)				
Iodine (I)				

Templates

Two Days a Week Schedule Template

Week: _____   Topic: _____

Day 1	Day 2
Supplies I Need for the Week	
Things I Need to Prepare	

Week: _____   Topic: _____

Day 1	Day 2
Supplies I Need for the Week	
Things I Need to Prepare	

Five Days a Week Schedule Template

Week: _____   Topic: _____

Day 1	Day 2	Day 3	Day 4	Day 5

Supplies I Need for the Week

Things I Need to Prepare

Week: _____   Topic: _____

Day 1	Day 2	Day 3	Day 4	Day 5

Supplies I Need for the Week

Things I Need to Prepare

Scientist Biography Report Grading Rubric

Spelling (points x 1)
- ✓ 4 points: No spelling mistakes.
- ✓ 3 points: 1-2 spelling mistakes and not distracting to the reader.
- ✓ 2 points: 3-4 spelling mistakes and somewhat distracting.
- ✓ 1 point: 5 spelling mistakes and somewhat distracting.
- ✓ 0 points: > 5 spelling mistakes and no proofreading obvious.

Points Earned _____

Grammar (points x 1)
- ✓ 4 points: No grammatical mistakes.
- ✓ 3 points: 1-2 grammatical mistakes and not distracting to the reader.
- ✓ 2 points: 3-4 grammatical mistakes and somewhat distracting.
- ✓ 1 point: 5 grammatical mistakes and somewhat distracting.
- ✓ 0 points: > 5 grammatical mistakes and no proofreading obvious.

Points Earned _____

Introduction to the Scientist (points x 2)
- ✓ 4 points: Includes thorough summary of the scientist's biographical information and why the student chose the particular scientist.
- ✓ 3 points: Adequate summary of the scientist's biographical information and why the student chose the particular scientist.
- ✓ 2 points: Inaccurate or incomplete summary of one of the scientist's biographical information and why the student chose the particular scientist.
- ✓ 1 point: Inaccurate or incomplete summary of both of the scientist's biographical information and why the student chose the particular scientist.
- ✓ 0 points: No introduction

Points Earned _____

Description of the Scientist's Education (points x 2)
- ✓ 4 points: Includes thorough summary of the scientist's education.
- ✓ 3 points: Adequate summary of the scientist's education.
- ✓ 2 points: Inaccurate or incomplete summary of one of the scientist's education.
- ✓ 1 point: Inaccurate or incomplete summary of both of the scientist's education.
- ✓ 0 points: No description of the scientist's education.

Points Earned _____

Description of the Scientist's Major Contributions (points x 2)
- ✓ 4 points: Includes thorough summary of the scientist's major contributions.
- ✓ 3 points: Adequate summary of the scientist's major contributions.
- ✓ 2 points: Inaccurate or incomplete summary of the scientist's major contributions.
- ✓ 1 point: Inaccurate and incomplete summary of the scientist's major contributions.

✓ 0 points: No description of the Scientist's Major Contributions and Interesting Facts of their life.

Points Earned _____

Conclusion (points x 2)
   ✓ 4 points: Explanation of why the student feels one should study the scientist and a summary statement about the scientist.
   ✓ 3 points: Adequate explanation of why the student feels one should study the scientist and a summary statement about the scientist.
   ✓ 2 points: Incomplete or incorrect explanation of why the student feels one should study the scientist and a summary statement about the scientist.
   ✓ 1 point: Conclusion does not have an explanation of why the student feels one should study the scientist and a summary statement about the scientist.
   ✓ 0 points: No conclusion.

Points Earned _____

Final Score = (Total Points/40) x 100%

Total Points Earned _____

Final Score _____

Periodic Table of the Elements

1	2	3	4	5	6	7	8	9	10	11	12	13	14	15	16	17	18
1 **H** Hydrogen 1.008																	2 **He** Helium 4.003
3 **Li** Lithium 6.941	4 **Be** Beryllium 9.012											5 **B** Boron 10.81	6 **C** Carbon 12.01	7 **N** Nitrogen 14.01	8 **O** Oxygen 16.00	9 **F** Fluorine 19.00	10 **Ne** Neon 20.18
11 **Na** Sodium 22.99	12 **Mg** Magnesium 24.31											13 **Al** Aluminum 26.98	14 **Si** Silicon 28.09	15 **P** Phosphorus 30.97	16 **S** Sulfur 32.07	17 **Cl** Chlorine 35.45	18 **Ar** Argon 39.95
19 **K** Potassium 39.10	20 **Ca** Calcium 40.08	21 **Sc** Scandium 44.96	22 **Ti** Titanium 47.87	23 **V** Vanadium 50.94	24 **Cr** Chromium 52.00	25 **Mn** Manganese 54.94	26 **Fe** Iron 55.85	27 **Co** Cobalt 58.93	28 **Ni** Nickel 58.69	29 **Cu** Copper 63.55	30 **Zn** Zinc 65.39	31 **Ga** Gallium 69.72	32 **Ge** Germanium 72.61	33 **As** Arsenic 74.92	34 **Se** Selenium 78.96	35 **Br** Bromine 79.90	36 **Kr** Krypton 83.80
37 **Rb** Rubidium 85.47	38 **Sr** Strontium 87.62	39 **Y** Yttrium 88.91	40 **Zr** Zirconium 91.22	41 **Nb** Niobium 92.91	42 **Mo** Molybdenum 95.94	43 **Tc** Technetium 98.91	44 **Ru** Ruthenium 101.1	45 **Rh** Rhodium 102.9	46 **Pd** Palladium 106.4	47 **Ag** Silver 107.9	48 **Cd** Cadmium 112.4	49 **In** Indium 114.8	50 **Sn** Tin 118.7	51 **Sb** Antimony 121.8	52 **Te** Tellurium 127.6	53 **I** Iodine 126.9	54 **Xe** Xenon 131.3
55 **Cs** Cesium 132.9	56 **Ba** Barium 137.3	*	72 **Hf** Hafnium 178.5	73 **Ta** Tantalum 181.0	74 **W** Tungsten 183.9	75 **Re** Rhenium 186.2	76 **Os** Osmium 190.2	77 **Ir** Iridium 192.2	78 **Pt** Platinum 195.1	79 **Au** Gold 197.0	80 **Hg** Mercury 200.6	81 **Tl** Thallium 204.4	82 **Pb** Lead 207.2	83 **Bi** Bismuth 209.0	84 **Po** Polonium [209]	85 **At** Astatine [210]	86 **Rn** Radon [222]
87 **Fr** Francium [223]	88 **Ra** Radium [226]	**	104 **Rf** Rutherfordium [261]	105 **Db** Dubnium [262]	106 **Sg** Seaborgium [266]	107 **Bh** Bohrium [264]	108 **Hs** Hassium [269]	109 **Mt** Meitnerium [268]	110 **Ds** Darmstadtium [272]	111 **Rg** Roentgenium [272]	112 **Cn** Copernicium [285]	113 **Nh** Nihonium [286]	114 **Fl** Flerovium [289]	115 **Mc** Moscovium [289]	116 **Lv** Livermorium [293]	117 **Ts** Tennessine [294]	118 **Og** Oganesson [294]

* Lanthanides

57 **La** Lanthanum 138.9	58 **Ce** Cerium 140.1	59 **Pr** Praseodymium 140.9	60 **Nd** Neodymium 144.2	61 **Pm** Promethium [145]	62 **Sm** Samarium 150.4	63 **Eu** Europium 152.0	64 **Gd** Gadolinium 157.3	65 **Tb** Terbium 158.9	66 **Dy** Dysprosium 162.5	67 **Ho** Holmium 164.9	68 **Er** Erbium 167.3	69 **Tm** Thulium 168.9	70 **Yb** Ytterbium 173.0	71 **Lu** Lutetium 175.0

** Actinides

89 **Ac** Actinium [227]	90 **Th** Thorium 232.0	91 **Pa** Protactinium 231.0	92 **U** Uranium 238.0	93 **Np** Neptunium [237]	94 **Pu** Plutonium [244]	95 **Am** Americium [243]	96 **Cm** Curium [247]	97 **Bk** Berkelium [247]	98 **Cf** Californium [251]	99 **Es** Einsteinium [252]	100 **Fm** Fermium [257]	101 **Md** Mendelevium [258]	102 **No** Nobelium [259]	103 **Lr** Lawrencium [262]